Flying Vehicles

The Emergence of Personal Air Travel, Flying Cars, and Air Taxis

Chuck Martin

Books by Chuck Martin

DIGITAL TRANSFORMATION 3.0: The New Business-to-Consumer Connections of The Internet of Things (NFI Publishing)

MOBILE INFLUENCE: The New Power of the Consumer (Palgrave MacMillan)

NET FUTURE: The seven cybertrends that will drive your business, create new wealth and define your future (McGraw-Hill)

THE THIRD SCREEN: Marketing to Your Customers in a World Gone Global (Nicholas Brealey)

THE DIGITAL ESTATE: Strategies for Competing, Surviving and Thriving in an Internetworked World (McGraw-Hill)

MANAGING FOR THE SHORT TERM: The New Rules for Running a Business in a Day-to-Day World (Doubleday)

TOUGH MANAGEMENT: The 7 Ways to Make Tough Decisions Easier, Deliver the Numbers, and Grow Business in Good Times and Bad (McGraw-Hill)

MAX-e-MARKETING IN THE NET FUTURE: The Imperatives for Outsmarting the Competition in the Net Economy (co-author) (McGraw-Hill)

THE SMARTPHONE HANDBOOK (NFI Publishing)

COFFEE AT LUNA'S: A business fable: Three Secrets to Knowledge, Self-Improvement, and Happiness in Your Work and Life (NFI Publishing)

Chuck Martin
Chuck@NetFutureInstitute.com

Published by NFI Publishing

10 9 8 7 6 5 4 3 2 1

ISBN: 978-0-9763273-4-9

For further information
603-770-3630
www.NetFutureinstitute.com
Printed in the United States of America
First Edition
Publication Date March 2024

Library of Congress Control Number: 2024902189
Library of Congress Cataloging-in Publication Data
Martin, Chuck, 1949-

Flying Vehicles: the emergence of personal air travel, flying cars, and air taxis/ chuck martin

p. cm.
Includes index.
ISBN: 978-0-9763273-4-9

1. Flying vehicles 2. Flying cars 3. Air taxis. 4. Technology 5. Aerial travel. 1. Title

To Madeline Marie,
Sky rider of the future

CONTENTS

Acknowledgements

This book about flying vehicles would not have been possible without all the innovators and developers who have been creating and testing the different approaches to personal flying vehicles, flying cars, and air taxis for many, many years. Their forward thinking, inventive spirit and tenacity in revolutionizing aerial transportation were the inspiration for this book.

Thank you to all those who agreed to be interviewed and freely share the stories of their journeys, both on and off the record. A special thanks to Sean Borman of Aeroauto Global, Doron Merdinger of Doroni Aerospace, Bob Johnson of Volatus Infrastructure, Santh Sathya of LuftCar, and Sam Bousfield of Samson Sky for participating in lengthy podcasts and discussions about the industry and their vehicles.

A special thank you to Mick Kowitz and the team at Ryse Aero Technologies, for providing a detailed tour of the facility, training me, and letting my fly the Ryse Recon, helping to understand the significance of the personal aerial vehicles detailed in this book.

Thanks to Lauren Prager and Basha Ferdinand at Synapse for recognizing the significance of the coming market of flying vehicles and helping aid market awareness by including it on the program at the annual Synapse innovation mega event in Tampa.

Thank you to colleagues at Informa Tech and especially to Liz Hughes, Editor of IoT World Today, for always grasping and supporting the import of emerging technology.

Mostly I want to thank my family for always being so supportive of my writing and all that it takes to complete a book, having witnessed it a number of times over the years. Thank you to my sons Ryan and Chase for always sharing their very tech savvy insights along with astute questions and challenges and my daughter in-law Kelsey for her patience and understanding that anything is possible.

And the biggest thank you goes to Teri, my wife and lifelong partner in everything, especially for the reality checks when I would tell her of the latest flying vehicle invention or innovations. And then the next, and the next. Thank you for the patience with me on this thirteenth book adventure, and for being all-in on all of them. Thanks for the constant questions, comments, and clarification suggestions all the way through the book development and mostly for being willing to learn and hear more than you ever wanted to know about flying vehicles.

1 INTRODUCTION

A New Era of Aerial Transportation

It's been a long time coming, but the electric aerial revolution is here. This new, plug-in, flying economy is about to transform personal and business travel, impacting logistics and supply chains in a coming fundamental shift in how people and things are transported. The new electric flying phenomenon is going to reinvent personal aerial travel with flying cars, air taxis, new first responder capabilities, and totally self-flying vehicles, enabling millions of consumers around the world to take to the skies just as they adopted road travel decades ago. This emerging technology is about transporting people and things totally differently than in the past and creating efficiencies in travel to locations previously out of practical reach.

Electric flying vehicles have been in development around the world for more than a decade with many thousands of miles of test flights already conducted. The new flying vehicles are electric and quiet with battery charging between flights. Flying vehicles have several separate motors, separate propellers, and separate battery systems, with safety factors incorporated so that systems are redundant. Unlike commercial aircraft that fly 30,000 to 40,000 feet, flying vehicles fly at much lower altitudes, in the 1,500-3,500 feet range. Flying vehicles are referred to as eVTOL vehicles, for electric vertical takeoff and landing, since the crafts take off and land straight up and straight down. They do not require a runway, though commercial operations will start at traditional airports and be integrated with current commercial air traffic. The first category of flying vehicles coming to market does not require a pilot's license to fly.

Unlike traditional airplanes, flying vehicles carry one to five passengers. They are designed for trips too short for a commercial flight and too long for a practical road trip. Commercial flying vehicles aim to significantly streamline travel. A trip from John F. Kennedy International Airport in New York to downtown Manhattan would transform a one-hour drive into a seven-minute flight. A New York City heliport is already being planned to handle eVTOL traffic.

Flying Vehicles Categories

Not all flying vehicles are the same. Some are designed to carry one person, some are designed to carry several people, some can drive on roads and convert to fly, and others are intended to be used as transportation vehicles where pilots can shuttle passengers back and forth between locations. There also are self-flying vehicles aimed at picking up passengers much like an Uber, albeit with no driver. The destination would be pre-programmed where to take the passenger, a category happening first in Asian markets. We detail each of the different categories more thoroughly later in the book, but here's how they break down:

- **PAV - Personal Aerial Vehicle**. This is defined as a powered ultralight by the Federal Aviation Administration (FAA). These are single-passenger vehicles that do not need a pilot license to fly. They can range in price from $95,000 to $300,000.
- **PAVC - Personal Aerial Vehicle Certified**. These are personal aerial vehicles that are FAA certified, meaning they conform to strict regulations and require a pilot license to fly. Vehicles designed to both legally drive on roads and fly also fall into this category. They can range in price from $300,000 to $1 million.
- **CAV – Commercial Aerial Vehicle**. Developed as people transportation vehicles, such as air taxis. This category includes electric jets and self-flying vehicles. This category is the most highly FAA regulated category, similar to the commercial airline industry. This is the category the military is experimenting with. These can range from $25 million to significantly higher amounts.

From a timing perspective, the PAV is coming first, with many shipping to buyers in 2026 and 2027. There are hundreds of pre-orders for these at numerous companies around the world. The general appeal is that they are relatively easy to learn to fly, since they typically are flown with two joysticks, one for going up and down and the other for heading directions. These ultralights also do not require a license, they are the least expensive of all the categories, and there is a large selection from which to choose. I flew one of these in the course of research for this book. The PAVC and the CAV for public transportation are not likely to be in operation in the United States until 2026, a target date set by the FAA along with numerous electric aerial vehicle makers.

Commercial flying vehicle operations will start with pilots on board, though the longer-term plan is to transition to pilot-free aerial travel. Many of the flying vehicles already have remote or autonomous flying capabilities and one flying vehicle company in China already is approved for carrying passengers in its self-flying vehicles. The Federal Aviation Administration has laid out a blueprint and initial rules for how these passenger-carry electric flying vehicles will come to market, though pilotless travel is not yet included.

The long-range implications of the electric aerial revolution are profound, with thousands of low-flying, electric aerial vehicles in the sky. Wide-ranging impacts include future road-based travel and all its components, the autonomous drone delivery of goods, the warehousing and distribution of products, improved emergency services, and new air ridesharing capabilities. The long-term future will impact all land travel, as much of moving around transitions to the air.

Air Roads

While autonomous or self-driving cars have been in development for many years, they still face the challenge of dealing with the legacy system of cars that are not as technologically equipped. Self-driving cars can do very well with other highly connected vehicles. It's all those millions of manually driven cars that can get in the way. Various forms of what I call air roads can be created in the sky with lanes designated for air taxis, personal flying machines, and other electric air vehicles not yet conceived. The air has no legacy systems to contend with.

Business Transformation

Flying vehicle are creating a low-altitude economy, which is being borne out of startups, established brands, and forward-looking governmental agencies fueling the velocity of aerial innovation. Billions of dollars already have been invested in flying vehicle companies and major airlines, automakers, and technology companies are all involved. There are even flying vehicle showrooms set up, such as those by Aeroauto Global in Florida with another on the way in Austin, Texas. The first vehicles to be delivered via showroom sales are expected to be personal aerial vehicles, the first of the flying vehicles to be delivered in 2026.

There are numerous PAV makers globally and many have been developing their flying vehicles for more than 10 years. PAVs are officially considered ultralight vehicles, with certain limitations required by the FAA, the federal agency that oversees all aerial traffic. For example, PAVs can only go so fast, have weight limitations, and can only carry one person. Some of these are expected to be used as recreational vehicles and others by landowners and farmers to quickly travel to a part of their land.

Certified Vehicles

Certified personal aerial vehicles, the PAVC category, will come later since they have to meet strict FAA rules for flight certification. Many of these are in the test phase, having received air worthiness clearance from the FAA allowing for flight tests. From a regulatory viewpoint, PAVCs are essentially viewed as passenger-carry airplanes, so the certification process is appropriately extensive.

Flying cars, vehicles that can drive on roads and then convert to fly also are coming, with several already conducting land and air trials, as detailed later in the book. A flying car from Chinese automaker XPeng Aeroht was featured on the floor of CES in Las Vegas in early 2024 where a Hyundai company introduced its four-passenger, piloted eVTOL with eight rotors. There are two different types of flying cars. There are those that comprise one complete vehicle that can quickly convert and take off into the air. The other flying car approach is modular, where the air vehicle separates from the road vehicle to fly. Both flying car types already exist and are in tests. The flying car is on the horizon and should be here before many people expect it.

While eVTOL vehicles do not require runways for takeoffs and landings, they still need a place to do that as well as have maintenance, such as battery charging between flights. That's where vertiports come in. Vertiports are for vertically operated vehicles, much like helipads are used for helicopters. There are numerous vertiport developments in the works globally, including in Europe, the Middle East, Asia, and the United States. Many of these are being set up to handle the wave of air taxis coming.

Flying taxis are a very large-scale proposition, causing an upheaval in how people move, leading to a transformation of where people go and how they get there. Most consumers are not likely to own a flying vehicle any time soon. However, they may find they have an opportunity to travel in one due to flying taxis, a key component of the low-altitude economy. Many air taxi vehicles are certified for flight testing and have been in tests and public demonstrations around the world. Major airlines are involved in air taxi businesses as extensions of their travel operations. The idea is an airline passenger could land at an airport and transfer to a flying vehicle for a short trip into the city, bypassing all the road traffic. For example, a flying taxi trip from Heathrow to downtown London would take 15 minutes.

Down the Road

The long-term future for flying vehicles is for them to be fully automated and fly without pilots. There is a blueprint for how this could occur by the end of the decade, which is detailed later in the book. But for the shorter term, the future of transportation is looking up with pilots on board.

There will be some skeptics relating to flying vehicles, a normal occurrence in technological revolutions. There have been skeptics for all new technologies with predictions of failure just around the corner. In 1996, Time Magazine predicted the failure of online shopping ("Remote shopping, while entirely feasible, will certainly flop. It has no chance of success.") and in 1992, technology superstar Intel CEO Andy Grove predicted the failure of the smart phone ("The idea of a personal communicator in every pocket is a pipe dream driven by greed.")

The electric aerial revolution is here. Just because someone doesn't see it doesn't mean it isn't happening. This revolution also involves the creation of a new business and economic climate, which we discuss in the next chapter.

2 THE LOW-ALTITUDE ECONOMY

The New Aerial Business

While it may take many in the general population by surprise, the electric aerial vehicle revolution has been years in the making. While some startup companies are but a few years old, others go back more than a decade. No matter the reactions, there is little doubt that a significant and substantial new market is arriving. But make no mistake, the electric aerial vehicle revolution is not only about startups, since established brands, including major airlines, global automakers, and technology powerhouses are highly active, some publicly and others behind the scenes. Investments in flying vehicle companies already reach into the billions of dollars and come from deep-pocketed, established brands, venture investors, and individuals who want to be part of the flying vehicle revolution in one way or another.

The short-term revenue will come from sales of personal aerial vehicles, since those are aimed at individuals and do not require a pilot's license to fly, as we detail in the next chapter. However, the larger revenue stream is a bit down the road as electric aerial vehicles are delivered to the airlines, air taxi companies, various branches of the military, tour operators, and general transportation companies. While the market is starting with the single-person PAV, vehicles in development that ultimately will be certified are being created to carry several passengers at a time as well as provide quick shuttle services. Revenue will be derived from numerous sources:

- Sales of aerial vehicles, some into the millions of dollars each, have already started, with major airlines and the military placing large bets.
- Flying taxi services, such as those planned at Dallas-Fort Worth Airport, in Dubai, at Tampa International Airport, from a heliport in New York City, and in Osaka, Japan.
- Touring by eVTOL, such as tours of Manhattan planned by Lift Aircraft from Texas. Shenzhen Boling Holding Group plans sightseeing tours in China's Guangdong province using the EHang autonomous aerial vehicle and sightseeing tours are being conducted in China in the company's flying vehicle. And Japanese eVTOL developer SkyDrive is selling vehicles to the MASC General Incorporated Association to conduct aerial tourism in Japan and AutoFlight delivered its five-passenger Prosperity to an advanced mobility operator in Japan to demonstrate EAV flights at the 2025 Osaka World Expo.
- Entertainment such as an indoor ride in a flying vehicle by Land Rotor, in Orlando, Florida, and

people flying vehicle through a geofenced course in the air created by Neo Aeronautics in Singapore.

- Emergency uses, such as the flying vehicles from Jump Aero, which can carry medical personnel and equipment, for use by the global Falk Ambulance Service and flying vehicles from Lift Aircraft in
- Texas for public agencies, such as medical and first responders.

While there is plenty of potential revenue ahead for the flying vehicle industry, there also are significant expenses. Developing a new form of flying, whether for individual and personal travel or the commercial transportation of passengers, is very costly. It takes staffing, software development, testing equipment, regulatory expertise, test pilots, and facilities. There also are no guarantees of success. Many companies in the first phase, the PAVs in the ultralight category, raised money from crowdfunding, self-funding, angel investors, and venture capitalists. Some early investors are likely to end up with windfalls while others will be on the other end. There also are some name investors, such as singer, songwriter, and technology entrepreneur Will.i.am, who was one of the early investors in the flying vehicle startup Jetson. An expect shakeout started in 2025 as companies continue to compete for investment dollars from a limited number of sources. Some companies are likely to team with others or even fully merge. However, many of the approaches to flying vehicles are dissimilar, as discussed later in the book.

Feeding the Ecosystem

The development of the flying vehicle industry is not without risks. To start with, the marketplace is both cooperative and competitive. Because there are so many different components, companies very naturally become part of the overall flying vehicle ecosystem. Manufacturers work with operators, operators coordinate with vertiport developers, vertiport developers team with electric charging system makers who also work with vehicle makers, manufacturers deal with flying vehicle sellers and distributors, and numerous entities work with the military. However, it is a relatively small community. While researching this book, it became clear that people in the industry closely track everything that is happening and that it is an emerging industry where 'everyone knows everyone.' I would be at a flight demonstration in a U.S. city speaking with a representative of the flying vehicle company and we would be discussing a test flight just conducted by a different company in another country. Or I would be doing a phone interview with a vertiport developer and mention a flying vehicle maker in another country and the vertiport developer would say that he just spoke with the founder of that company the day before. Because the industry is still in the emerging stage, it is in the interest of everyone involved to contribute to the success of the overall market.

The Plug-In Economy

The new flying vehicle industry is primarily electric. In addition to being relatively quiet, the vehicles need charging. Unlike the number of charging stations needed for the millions of cars on the road, fewer setups are required for flying vehicles, at least at the start. Stations are being installed at airports where electric aerial vehicles are planned to land or take off. Unlike cars, a flying vehicle with wings or protruding propellers can't just drive up to a charging station and plug in. Longer cables to reach the vehicles are required along with higher power requirements. Some charging stations being installed at airports also can be used to charge land vehicles, thereby increasing efficiencies. And in true flying vehicle ecosystem form, companies work together. For example, Beta Technologies partnered with Archer Aviation to develop interoperable charging stations in numerous locations and already had charging stations in more than a dozen locations in the eastern United States, including Eglin Air Force Base in Florida, where the U.S. Air Force is testing one of Beta's electric aerial vehicles. Beta flying vehicle charging can be done in less than one hour. Charging procedures at vertiports as well as charging systems are expected to be standardized over time as the industry evolves, grows, and matures.

The flying vehicle industry, years in the making, has had many companies involved, although it was not highly publicized in the past as the market developed. Companies in various industries are always evaluating emerging technologies, partly for competing reasons and partly for new opportunities, either in new revenue or cost and production efficiencies. This is true of the airline, automotive, and technology market segments.

The Airlines Move In

- **United Airlines** placed a $1.5 billion order for electric vertical takeoff and landing vehicles from Archer Aviation, a Palo Alto, California-based startup. United has an option to purchase another $500 million of Archer vehicles.

- **Delta Air Lines** made $60 million equity investment in Joby Aviation with a total investment of up to $200 million as further milestones are achieved. The partnership plans to deliver a premium service for Delta customers alongside Joby's standard airport service.
- **American Airlines** invested in Vertical Aerospace, a United Kingdon engineering and aeronautical business developing electric vertical takeoff and landing vehicles, in 2021. As part of the deal, American Airlines agreed to pre-order up to 250 aircraft, representing a potential pre-order commitment of $1 billion, and an option to order an additional 100 flying vehicles. The airline also expects to make a $25 million investment in Vertical Aerospace through a private investment in a public equity transaction.
- **JetBlue** Ventures JTV's invested in Joby Aviation in 2017 to align with its commitment to identify and invest in sustainable travel technology.
- **Embraer** is building an electric vertical takeoff and landing aircraft production facility in São Paulo, Brazil. The plant to build the Eve Air Mobility flying vehicles is situated on land within Embraer's existing unit in the city that will be expanded. Eve Air introduced a full-scale prototype of its EAV in May 2024 with pre-orders for nearly 3,000 vehicles.
- **Virgin Atlantic** and **American Airlines** signed two separate purchase deals with Vertical Aerospace, under which both companies will acquire up to 400 electric vertical take-offs and landing aircraft combined. Virgin Atlantic committed to purchasing up to 150 fully electric zero-emissions VA-X4 aircraft capable of

flying over 100 miles while carrying four passengers onboard.

- **Japan Airlines (JAL)** is working with Wisk Aero to bring self-flying, all-electric air taxi services to Japan. Wisk and JAL Engineering plan to work closely together to develop plans for the maintenance and operation of the Wisk autonomous air taxis.

The Automakers Move In

- **Hyundai Motor Group** created a company called Supernal in 2021. Supernal is developing a family of electric air vehicles with plans to launch a commercial service. The company even built a vertiport at CES 2024 in Las Vegas, where it showcased its flying vehicle prototype, though by late 2025 had put the project on hold.
- **Stellantis** is providing $150 million in equity capital for Archer Aviation with Stellantis becoming the exclusive contract manufacturer to mass produce the Archer Aviation Midnight flying vehicle. The two companies have been working together since 2021.
- A **Suzuki** company plant plans to manufacture the vehicles of SkyDrive, a Japanese flying vehicle maker. Suzuki and SkyDrive teamed in 2022 to work together on research and development, manufacturing, and mass production systems. Up to 100 of the SkyDrive vehicles a year could be created at the new subsidiary Sky Works Inc. at the Suzuki plant.
- **Honda** is working to create electric vertical take-off and landing aircraft and striving to establish a mobility ecosystem.

- **XPeng**, a Chinese automaker, established XPeng Aeroht, which introduced a modular flying car concept along with a vehicle described as like a moon rover for Earth. In late 2025, the company changed its name to Aridge.
- **Toyota** invested $400 million in Joby Aviation, making it Joby's largest external shareholder.

Powering and Funding

- **Honeywell** secured contracts worth $10 billion through deals with several electric aerial vehicle companies including Archer Aviation, Hyundai's Supernal, Eve Air Mobility, Pipistrel, and Vertical Aerospace. Honeywell provides the fly-by-wire technology and custom-built actuation systems, and its technology provides precise flight controls and redundancy for advanced vehicles. Honeywell sensors also provide attitude heading reference systems, making the vehicles aware of their environment.
- **Boeing** invested $450 million in self-flying air taxi developer Wisk Aero, which ultimately became a wholly owned subsidiary of Boeing.
- **Archer Aviation** raised a $215 million investment from companies including Stellantis, Boeing, United Airlines, and ARK Investment Management LLC, increasing Archer's funding to more around $2 billion.
- **Jetson**, a personal aerial vehicle maker, raised $15 million in financing from angel investors including singer, songwriter, and technology entrepreneur Will.i.am.

Unplugging

For some companies, flying vehicles are core to their business. For example, the entire goal and mission of EAV startups is to either sell the flying vehicles or the components to create and service them. For airlines, transportation is their core business, so flying vehicles can fit within their purview. Commercial aircraft manufacturers also can have a role here. For example, in 2024 Airbus introduced its EAV prototype CityAirbus NextGen, a four-seat flying vehicle with a 50-mile range, with flight testing planned for later in 2024. For automakers, in addition to being in the transportation business, the manufacture of flying vehicles can nicely fit within their manufacturing expertise and capabilities. However, for some companies, flying vehicles could fall outside their core businesses or not provide enough revenue compared to their central offerings.

Rolls-Royce technology was set to power the Vertical Aerospace flagship urban air mobility craft. A Rolls-Royce electrical power system was going to be integrated into the piloted all-electric vertical take-off and landing vehicle. The company had spent years working on advanced electric propulsion systems for flying vehicles. However, the Rolls-Royce central mission revolves around electrical engineering activities in Power Systems, Defense and Civil Aerospace. Rolls-Royce ultimately decided to withdraw from the electric flying business and stay focused on its jet propulsion business. The company put its electrical division up for sale in late 2023 and started negotiations with a potential buyer. Some companies go all-in on electric aerial vehicles, one example being Boeing acquiring Wisk Aero. Others, like Rolls-Royce and Hyundai, decide it is not needed in their core portfolio.

Flying Vehicle Showrooms

With so many flying vehicles coming into the market, one startup decided that centralized locations would be needed for people and businesses to see numerous vehicles, both for educational experiences and for the ultimate selection and purchase. The idea was to aggregate as many flying vehicle companies as possible and arrange to have those vehicles displayed in showrooms once they were ready. That startup was Aeroauto Global Inc., now headquartered in Royal Palm Beach, a village 15 miles west of Palm Beach and the Atlantic Ocean. Sean Borman is President and CEO of Aeroauto and he is creating fully integrated showroom dealerships specializing in aerial vehicle transportation, including complete sales and service of the vehicles.

"Aeroauto was born out of the idea of trying to find my second career," says Borman. "I was doing some searching and saw that the industry was coming, through Facebook and through Instagram, and a few articles here and there that just showed the flying cars were well on the way. I've always tried to get into an industry at the ground level, and I keep missing it for one reason or another. I saw that it was coming, and I had dinner with my partner and co-founder one night and I said flying cars are coming, we should do something like invent a part or an accessory or get into racing or something. He suggested doing a dealership, which was just so far outside of my scope of thinking at that point. I went home and started doing a lot of research and realized that there was no other company globally that was in the sales and service end of it, except for the manufacturers that were trying to do a Tesla model on their own, which is just trying to sell and create at the same time. That's a business model that's hard to pull off unless you're Elon Musk."

Early to Market

"We started a couple of years ago and that was really early. July 2023 was our two-year anniversary, but the industry is just starting to explode now, two years later. We wanted to be the first in. We've been working to land the manufacturers that are ready or starting to take reservations or make sales. We are trying to jump on them and to get them into our portfolio.

"We just did a deal with Ace VTOL in Perth, Australia. (Ace VTOL is featured in a later chapter). CEO Brett Northey is fantastic. He's got an amazing vehicle that's a hydrogen plasma ion powered. He's really ahead of the curve as far as technology goes and we're excited to see all the different vehicles he comes up with. Every time I turn around, he sends me a new version or a new vehicle that he wants to do later, in phase three or phase four. He's so far above and beyond all the other designers out there, as far as creativity goes."

The market challenge Borman identified was the lack of general knowledge about flying vehicles, especially since there are so many different varieties. The flying vehicle industry tends to be somewhat localized, at least in terms of PAVs and PAVCs. For example, the seller of a particular ultralight may focus marketing efforts near their main base of operations for practical reasons, such as costs and resources. Regionalization also allows for local training and vehicle servicing of that particular craft. Residents in that region may only become aware of that flying vehicle, due to proximity, leaving the overall market out of consideration.

"If someone walks into a showroom, there are different flying vehicles there," says Borman. "That's our business model. The whole point and purpose of what we're doing is customer education. By doing that, a customer comes in and has no knowledge or expectations or experience of what any of these flying machines can do other than the current helicopters and planes that are out there.

"There's no brand recognition and no option knowledge. There's no idea of what they can get or what they can't get with the price ranges and what their use cases are and could be. They come to us, and they say they have a budget of this and tell us their use case, and we then direct them to the appropriate vehicles for what they need."

Serving the Global Market

The flying vehicle industry is a global phenomenon, as discussed earlier. This makes it even more difficult for an individual or company to get a first-hand view of any number of flying vehicles, a void Aeroauto is looking to fill. In addition, the market continues to expand as new entrants develop more vehicles.

"The biggest problem right now is that all these vehicles are being distributed and created around the world," says Borman. "If there's no way to test drive these and to actually see one up close, you've got to go to California, then you've got to go to Japan and then go to Germany, and then to Florida. It just doesn't make any sense. We're trying to put as many dealers and manufacturers as possible into one place at a time to give the customer an opportunity to see, touch, feel, and experience these vehicles firsthand because they don't even believe that it's real, whether it's a customer off the street, a municipality, or a government agency. They can't even wrap their heads around the fact that these things are coming.

"We're trying to get as many flying vehicles in as we can. It takes a tremendous amount of money to go from concept to production and not all of them are going to pass their certifications. Not all of them are going to be able to find the funding, so we don't want to put all of our eggs in one basket. These are all brand new and untried. There are so many different makes, models, shapes, sizes, and use cases that are going to be available. We want to show the customer as many options as we can. We're the only one worldwide right now."

Showroom for Austin

"We're currently located at North Perry Airport near Hollywood, Florida, and at North County Airport in Palm Beach Gardens, Florida, says Borman. "We also just signed a big deal to open a large showroom in Texas at Greenport International Airport along with Emerald Island outside of Austin, Texas. We're very excited about that. We should have that huge facility open within the next two years.

"We want to have a lot of training for you before you get into it and take it home. The use cases are tremendous, all the way up to air cargo, large passenger, air taxi, municipal vehicles, police, fire, air ambulance, organ transportation, public works, infrastructure, and management vehicles. Search and rescue is going to be a huge opportunity for the Coast Guard, for lifeguards, forest and mountain management, and rescue. We're excited to keep seeing all the new use cases that are invented by the end users. Ambulance companies and police departments are going to want these.

"There are about 350 different manufacturers around the world, with close to 800 types of vehicles being developed. It's kind of a back and forth between the manufacturers that want the representation that Aeroauto is giving and the marketing, sales, and services that we provide to the ones that are available. We're going after them as they come to the market, as they start to show that they're available. But some of the companies want to maintain everything in-house from sales to maintenance, but that is very hard. If it was a sustainable business model, Chevrolet, Ford, Toyota, and Honda would all be owning their own dealerships instead of outsourcing it to the individual dealers around the world. They can't scale it, basically. They'll probably figure it out eventually and then they'll come talking to us. We'll have to say no at some point, but at the same time, the manufacturers are going to start going by the wayside or they'll start teaming up."

Low Air Travel

The idea of congestion in the sky due to the wave of EAVs coming can be significantly overstated. Just as commercial aircraft takeoff and land and travel around the globe without incident, the same can be expected with electric vehicles traveling in low-altitude airspace. Sensor and communication technology is advancing to manage the flow and low-air traffic.

"Within the next 20 to 25 years, it won't look any different than if you go outside and look up today," says Borman. "That seems to be a misunderstanding by the population, and I get this all the time: 'oh, there's going to be tons of people running into each other in the sky and falling out of the sky.' My response to them is there already are millions and millions of air vehicles in existence today. And very rarely do you ever hear about something colliding in midair and the technology that's coming into these vehicles today is just so much more advanced than ever been before, but they're pretty much low altitude.

"The FAA has been making a lot of decisions as far as the corridors that they're going to be flying in and it looks like they're going to stay in the helicopter and small plane levels of flight, so that's relatively low. None of these vehicles are pressurized to go that high anyway. If I go outside today, I may see one or two Cessnas flying across the sky, but certainly not enough to call it a traffic jam.

"Technology today is finally catching up to the desires and the requirements that are needed for a decent, sustainable flight. Electric batteries are light enough and quiet enough that now they can get you 20 minutes or an hour or two hours of flight versus a few years ago with battery technology giving you maybe five minutes. It just wasn't a good enough idea yet, but it's going now."

A Global Picture

Aeroauto deals with flying vehicle makers from all over the world and it is ready to display and sell a wide range of vehicles, including those from EHang in China, Ace VTOL in Australia, LuftCar in Florida, and Volocopter in Germany. As a result, Borman has to keep an eye on aerial vehicle developments globally.

"China is way ahead," says Borman. "It's unfortunate, the political piece of it, but the whole Asian area has adopted this technology significantly and allowed the development to excel, but the United States is catching up quickly. We have so many manufacturers here and the resources here are tremendous. Companies like Joby and Archer are making a big hit, but they just want to stay in the air taxi market and are forgetting all the other use cases that are possible from personal use and the municipal and first responder opportunities. We just want to expand. We announced a month ago that we signed on with another strategic partner and we now have reach into more than 100 airports coast to coast. We are reaching out to all those different areas, and we now have a distribution network and maintenance network from east to west, north to south. We want to be a specialized parts dealer along with it. Unfortunately, none of the companies are ready at that point yet, but as soon as they do, we'll start opening up a whole parts department, being able to stock all the parts and pieces necessary for all the maintenance. The nice part about this industry is the low maintenance that's required and the low number of parts that are required. A lot of it's going to be plug and play. Nobody has to take apart a gas engine anymore. You just take out the electric motor and put in a new one and ship the old one back to the manufacturer and let them deal with it."

The Piloting

While some leaders of flying vehicle companies have a private pilot's license, many do not. To operate a PAVC, a pilot license will be required, unless the FAA ultimately adopts different requirements of the category. From a regulatory standpoint, the final details of what the corridors in the sky will look like are still a work in progress, with the FAA working with flying vehicle developers and the public as it defines the new rules. Meanwhile, personal aerial vehicles that operate under the ultralight category of rules, with no pilot's license needed, are destined to be the first major wave in 2024. In the case of Aeroauto, there are some pilots on board, though Borman is not.

"I don't know if I should tell you the truth," says Borman during our discussion. "I kind of wish you hadn't asked this question, but the truth is, I'm afraid of heights. I don't fly very well. I can get into a 747, but don't put me in anything else. All my partners are pilots. The big joke is that somebody has to stay on the ground, take pictures and be the cheerleader, so that's become me."

Borman also expects the PAV category to be first to market "because they don't have to go through the FAA certifications. They don't fall under the same rules and regulations of the requirements necessary that the FAA has. They have a very specific range of speed and height and weight that they must maintain. We can start selling those today and all the other companies that are going from the light sport aviation companies to the larger air cargo, air taxi, and multi passenger manufacturers have to go through a lot more scrutiny through the FAA. The air taxis have to go through a whole other level of certification. Then there are the companies that want to do an autonomous only model. I haven't been able to figure out what their business model is and why they want to do that. They're going to be the last ones to be certified and there's going to have to be a lot of years of proof and trust, not just by the FAA but by the public."

Technology in the Sky

Borman sees the potential industry risks ahead but also understands the scope of advanced technology involved. He also sees the opportunity in new air travel and the safety capabilities that can come with the new flying vehicles.

"Aeroauto takes safety to a whole other level and understands that this whole eVTOL industry has a zero-tolerance opportunity and not just the FAA, but for the public itself," says Borman. "As soon as there's a crash, or if there's a death, the whole industry is done. However, what's beautiful about all the vehicles that we explore is the safety redundancies being put forth, whether it's multiple motors, multiple computers, ballistic parachutes, the ability to fly with less than the number of propellers and motors that they have. GPS is at a whole new level, lidar, and radar are at a whole new level. I've been telling our customers that you're flying in a bubble, and you'll never have anybody inside of your bubble and the sensors that are on these things will never let you crash into a building or a tree or another air vehicle. We're really excited about what's out there.

"Unlike ground transportation, in the air there are no bikes and kids and trees and fire hydrants and rocks and boulders and ditches and anything else that you're going to run into with a car. The sensors are going to be quiet until you say I want to land and then they start picking up the things around you. In the air, you're talking 100 years before there are enough vehicles in the sky to have any concern that you'd be running into anything. By then, all the sensors and data being able to talk to each other are going to be exponentially better."

Selling the Dream

The Aeroauto showrooms are where people can see and experience the flying vehicle future, giving Borman a front-row seat. Unlike a specific flying vehicle developer. Aeroauto gets to deal with them all, providing a global and horizontal view of the entire market and ecosystem.

"Aeroauto gets to sell the one thing that everybody's always wanted," says Borman. "I haven't had anybody come by that hasn't been full of smiles and awe and wonder. And even the people who have put down deposits on the vehicle, when they finally got to see it -- they bought it sight unseen or off a picture and they're like -- 'oh my God, this is real. I didn't even think this was going to be real.' It didn't matter who they are in the community, whether they're the upper echelon or the lowest echelon, male, female, and it doesn't matter from what country they're from. Everybody's always wanted what we're selling. We are at the forefront of this brand-new ecosystem that is every day being dreamt up, created, invented. It's like 150 years ago, when it was just a horse and buggy. When the first horseless carriages came out, there were no stoplights and parking lots and parking garages and gas stations and drive through restaurants and superhighways.

"All of this has come either through invention or the need and the desire to do new things and create new opportunities, whether it's because of traffic or because of the use case. Here we are with this brand-new opportunity of transportation that will become a sustainable and an ecologically better opportunity than driving to get us to where we have to go, both safer and quicker. I get to watch from ground level and have all these amazing designers and engineers and mechanics and architects around me that are dreaming up the newest, latest, and most amazing ecosystem that every day is going to be new. People are banging down the door. Everybody wants to be a part of this. Right now, we're primarily taking pre-production deposits and we've had quite a few. We've had about 50 in the last couple of weeks.

"The ultralights will be here soon. They're being done and processed and delivered as we speak. All the rest that aren't ultralights are still in the middle of their certification processes. I don't envy the manufacturers for that piece of it. We go to them and say, 'you guys go do what you do best, which is design vehicles, and let us take care of your customers and all the other problems that you don't want to deal with.'"

The new low-altitude economy is about the end-to-end flying vehicle revolution. It involves new revenue streams from sales, air taxi services, touring by eVTOL, and entertainment. An entire ecosystem is being created around flying vehicles. People are being hired to work in all parts of the ecosystem, including in growing startups, new segments in established companies, and in building and working in new facilities needed to build the coming fleets. More new companies will enter the market while others leave, either by merger, acquisition, or by running out of runway to financially operate. Major entities including airlines, automakers, and technology companies will get involved or increase activity. Showrooms will be opened around the world for the world to see the flying vehicle future up close. Individuals will finally get to see the personal air travel of the future, which we discuss in the next chapter.

3 PAV

The Personal Aerial Vehicle

The personal aerial vehicle – the PAV—comprises the first phase of the electric aerial vehicle revolution, for several reasons. Compared to certified personal aerial vehicles (PAVC) and commercial aerial vehicles (CAV), the PAV is relatively straightforward to develop. There are lower barriers to market entry in terms of selling opportunities, since the PAV is generally available at a lower price point, though still higher than the average price of a new car. Prices eventually should come down as components start to be produced at scale. Flying the vehicle also does not require a pilot license, although the buyer takes on certain liabilities, which we discuss later in this chapter. There are also numerous PAV makers, providing multiple choices for potential buyers with competitive market dynamics.

Years in the Making

PAVs have been in development for years with some startups developing their flying vehicles for up to 10 years. Many test and demonstration flights occurred throughout 2023 in different countries. A PAV developer typically creates a testing prototype vehicle, with many tests conducted without anyone in the vehicles. Like drones, a PAV often can be controlled remotely from the ground, which allows for early testing while limiting personal risks. While a prototype PAV is being flown, tested, and demonstrated, learnings from all those experiences are being incorporated into the next version, since the companies all are working toward production versions after the prototype and testing phase. While some leaders of PAV companies have a private pilot license, they are not required. Many of the vehicles are shipping to buyers in 2024, the first category of the new electric flying vehicles. Many of the PAV makers have been taking pre-orders for more than a year.

The general public has been gradually hearing about PAVs from the developers as many of them invited investments in the form of crowdfunding, providing varying incentives. Several companies also have publicly demonstrated their flying vehicles. For example, Ryse Aero Technologies took one of its flying prototypes to CES in Las Vegas in 2023 and invited numerous people to fly it and Lift Aircraft introduced its Hexa PAV at SXSW in 2019. The PAV is essentially the introduction to the public market as a first glimpse at future electric aerial movement.

Rules to Fly a PAV

Personal aerial vehicles are officially considered ultralight vehicles, the same category as hang gliders, though with additional regulations due to being powered. PAVs must follow the Federal Aviation Administration regulations called FAA Part 103 (see Appendix B for the full FAA Part 103). Many of the rules have to do with limiting the capabilities and operations of PAVs. For example, a PAV may not be capable of traveling faster than 63 miles per hour and cannot weigh more than 253 pounds empty. The vehicles can be for only one person. PAVs are specifically not required to meet the airworthiness certification standards for aircraft or to have certificates of air worthiness. These types of certifications are required for certified personal aerial vehicles (PAVC), which is discussed later in the book. People flying PAVs are not required to meet any aeronautical knowledge or experience requirements to operate the vehicles. However, PAV makers typically provide advanced training before anyone flies any of their vehicles, both for safety reasons and to ensure the person has a positive experience.

PAV Operations

Personal aerial vehicles generally are flown by two joysticks, with one making the vehicle go up and down and the other making it go forward, backward, or turning. FAA Part 103 mandates that the vehicles can be flown only in daylight and in uncontrolled airspace. The most important element relating to flying the vehicle relates to liability residing with the operator and not the manufacturer or PAV seller. An FAA Advisory relating to Part 103 makes it clear who is responsible for operating a PAV (see Appendix C for the key portion of the informative FAA Advisory). "The FAA realizes that it is possible to design an ultralight which, on paper, meets the requirements of (Part) 103, but in reality, does not," states the FAA. "However, the designers, manufacturers of the kits, and builders are not responsible to the FAA for meeting those requirements. Operators of ultralights should bear in mind that they are responsible for meeting FAA Part 103 during each flight. The FAA will hold the operator of a given flight responsible if it is later determined that the ultralight did not meet the applicability for operations under Part 103." While PAV developers are aware of the FAA regulations, it's also in their long-term business interest to assure that their customers have a positive PAV experience. Every PAV developer we spoke to in the course of researching this book had an intense focus and emphasis on safety in all their development efforts. Systems in PAVs generally are fully redundant. For example, if one battery should fail, other batteries can remain in operation. Same with propellers. The vehicles also fly in the hundreds rather than thousands of feet.

PAV Uses

Personal aerial vehicles can be used as sport or recreational vehicles. Some of them can fit into the back of a pickup truck and be taken to rural or open areas to fly. They also could be used for short trips. For example, a PAV could be useful on a large farm where a farmer may want to check on a certain portion of a field without having to drive and walk the distance. Someone at a summer home on a lake could use a PAV to get to the other side without a boat or a rancher or landowner could check on large tracts of land. Military uses are also being considered for all the categories of electric aerial vehicles, as we discuss later in the book. With current battery technology, a PAV can fly for 20 to 30 minutes before needing to either charge or change batteries. That is expected to change as battery technology continues to improve, which PAV developers can account for in their vehicles designs. There are numerous developers of PAVs, so here are some of those in the lead.

Lift Hexa

Lift Aircraft in Austin, Texas, was founded in 2017 and introduced the Lift Hexa to the public in late 2018, following unmanned test flights and manned test flights earlier in the year. The vehicle was displayed at the well-known SXSW event in 2019. The vehicle has been featured on the Today Show and 60 minutes and in Dubai. Production began in 2019, was awarded an Air Force research and development contract with the U.S. Air Force for an air ambulance in 2020 and received a $2.5 million Air Force flight test contract. The first production airframes for vehicles, which can also land on water, were shipped in 2020. Lift is developing a vertiport in Austin. The Lift Hexa also has completed piloted flights in Japan and marketed rides to the public at the annal Sun 'N Fun Aerospace Expo in Lakeland, Florida. The Founder's Series Hexa is priced at $495,000.

Jetson One

Jetson was founded as Jetson Aero in Poland in 2017 and then moved to Sweden with the shortened name of Jetson. The Jetson One proof-of-concept prototype was completed in 2018. Production facilities are in Arezzo, , although the vehicle has been flown and will be sold in the United States. Jetson raised $15 million in seed funding from angel investors including singer, songwriter, and technology entrepreneur Will.i.am. The Jetson One has eight motors and a flight time of about 20 minutes. The entire 2025 production was sold out. The Jetson One is priced at $128,000. The first vehicle was delivered to a buyer in 2025.

Land Rotor Sportster

Land Rotor, in Orlando, Florida, planned to make its Sportster PAV prototype available first as an indoor ride with the vehicle tethered to the ground in a drone ride concept. Land Rotor intended to capture and analyze data from simulated rides. The company wanted to deliver up to 10,000 Sportster vehicles for personal use by 2030 through the Aeroauto, Florida, dealership. Priced at approximately $70,000.

Dragon Personal Air Vehicle

Rotor X Aircraft in Chandler, Arizona, maker of the Dragon Personal Air Vehicle, also makes a two-seat helicopter. Dragon test flights were conducted in 2022 and 2023. The vehicle comes in a kit to be assembled by the owner or a building assist program is offered, which includes instructions to fly the vehicle. Flight time is about 20 minutes and the batteries have a two-hour charge time. Development, internal funding, and testing of the Dragon was by Advanced Tactics and is licensed by Rotor X Aircraft to sell and produce the vehicle kits. The full kit for assembly is priced at $89,500.

Pivotal Helix

Pivotal, in Palo Alto, California, was originally known as Opener, and started developing its flying vehicles in 2011, with a pre-production version making its first proof-of-concept flight in that year. The pre-production version called BlackFly was first flown in 2014 with its first manned flight in 2018. By 2019, the vehicle had flown 30,000 test miles. In late 2023, Pivotal introduced the Helix, the company's first scalable production vehicle. The tilt aircraft has eight fixed rotors and tandem wings, can emergency land on water, and fly with maximum wind of 20 miles per hour. Pivotal CEO Ken Karklin featured the vehicle at CES 2024 and CES 2025 in Las Vegas. The Helix went on sale in early 2024 and is priced at $190,000.

Zapata Airscooter

Zapata in France has been developing air vehicles since 2016 and introduced a prototype of the Zapata Airscooter in 2023. Rather than a full electric system, the Airscooter has a hybrid propulsion system that combines thermal and electrical energy, suggesting a flight time of up to two hours. The multi-copter style PAV is an egg-shaped pod. The Airscooter is not yet priced.

Neo Aeronautics

Neo Aeronautics in Singapore is an eVTOL startup that markets EAV flying experiences in its single-seater Crimson SS vehicle. Neo is working with the Oklahoma Department of Commerce to bring automated flying vehicles to the state to boost tourism. People would fly in the vehicle through a geofenced course in the air. Unlike the others, this one is not planned for individual sales at the moment but could provide the ability for someone to take a ride to see how they like it. At the Neo exhibit at the DriftX Mobility Expo in Abu Dhabi, CEO Neo Kok Beng told me his coming vehicle would be like "a go-cart in the sky."

Ryse Aero Technologies

Ryse Aero Technologies, headquartered in Mason, Ohio, was founded in 2021. The company's flagship flying vehicle is the Ryse Recon, which was included in Time's Best Inventions of 2023 in the Transportation category. One of the target markets for the Ryse Recon is to help ranchers and farmers quickly get to remote parts of their land. Mick Kowitz is the founder and CEO of Ryse Aero Technologies, with a background in technology and artificial intelligence. I met with Kowitz and the Ryse Aero team during an extensive visit and tour of the Ohio headquarters.

As with many of the company executives I spoke with in researching this book, personal motivation is a factor behind the desire to start developing flying vehicles of one type or another. "It's kind of a passion project to some degree as I think most people in the space are," says Kowitz. "I also happen to be a private pilot, so that factors into it. I've been flying since I was 15, so that obviously is a big part of it. I've been involved in several projects that have been close to this, but this was one where I thought, with the venture capital firm we're working with, this had real legs. I think you have to have a little something where you have a passion for it. It's different than just building software systems, of course, coming out of artificial intelligence and natural language processing in the healthcare world where you're dealing with lives, and you've got to make sure that you're protecting the patient's rights. It's kind of similar."

As a licensed private pilot, Kowitz has a deep understanding what is involved in flying different categories of flying vehicles. While flyers of the powered ultralight category of flying vehicles do not need any formal training or licensing, pilots of certified flying vehicles, the PACV category, require extensive ground and flight training before licensing.

"To become a private pilot or somebody who just flies a regular everyday aircraft, it takes a lot of effort," says Kowitz. "The skillset is deep and wide. You have to have a lot of troubleshooting skills, and aircraft that have airworthiness certificates are very complex from the perspective of having all the different air worthiness protections that are out there that the FAA formulates. With this, we're really an ultralight aircraft. What we wanted to do was make it possible for anyone to fly without a pilot's license. Some of the other players in the market are doing personal aircraft, but they're actually going to fall under the FAA part where you're going to have to be a pilot to fly those. For this, you do not have to be a pilot. It has its limitations. It can't weigh too much it has to be a single occupant. You can't fly over congested areas and has a limitation of speed."

Safety Training

While the FAA does not require any formal training for someone to fly a powered ultralight, Ryse Aero and other PAV makers typically provide and even require training. "I could teach you to fly this aircraft in 15 minutes," says Kowitz. I went through the training with Kowitz before getting into the flying vehicle for a test flight, which mostly focused on safety issues and emergency procedures if something were to go wrong. Kowitz described the details of the company's PAV.

"It's called the recon, the Ryse Recon. It actually can land and take off in water. It has a six independent propulsion module system, open cockpit with two joystick controls. It takes off and lands in water and it has removable rechargeable batteries. The vehicle is designed very simplistic like an EV lawnmower, but with six independent propulsion modules that are away from the operator. Many configurations have the battery underneath where the operator sits. This has the batteries disjointed, distributed, if you will, out at each pod. They each have their own battery management system and their own speed controller.

"We do have redundant flight control systems and redundant tertiary redundancy level of communications across all the pods. But if one pod stops working, completely disconnected from everything else, the vehicle continues to fly safely, and you land safely. If the battery stops working in one of those pods, all six motors will continue to work because not only are they distributed, but they are also tied together, so that it can help that pod continue to fly. We've really built a level of redundancy that means if everything goes out, you're not going to just fall out of the sky. It will safely get you to the ground no matter what."

Most PAV's are not designed to land and takeoff on water, although some account for emergency water landings. The Ryse Recon, however, was initially designed for on and off water operations. Ryse Aero also produced a video showing the vehicle landing on a lake, floating while the motors were powered off, and then started and taking off from the lake. "Each outrigger, each of the six propulsion modules, has a float," says Kowitz. "It has an outrigger float, and then underneath the operator is actually an airflow, so that is enough dissipation of water in order to be able to keep you from sinking. You can go in and land."

Powered ultralights can have various uses, from recreational to practical. After many market tests, Ryse Aero zeroed in on a specific market, which involves more practical, short-trip transportation outside of recreational use.

"We wanted to pick a target market to begin with that has a meaningful purpose to using the vehicle," says Kowitz. "We started going to farm shows. We saw agriculture as a very good first entry, for a couple reasons. From the sales perspective, you have the owner, the operator, and the check signer. All in one is the person who buys it. Agricultural people have adopted technology earlier than almost anybody. They were first to use stitched imagery, first to use GPS satellite, first to use drones, first to use autonomous vehicles. And what other industry touches every consumer in America three times a day. So they're a great market. They spend $54 billion a year on agricultural large equipment. Your average farmer spends about $650, 000 a year on farm equipment. There are a lot of farms so that's a lot of people and a lot of opportunity. And they respect equipment. They understand safety and they understand how to use it.

"They can use it for multiple purposes. One, to be able to go as the crow flies and get across their fields, where right now they have to drive zigzag around them. Crop scouting, to be able to get out to the center of an irrigation system instead of walking a mile. They can land right out there with the equipment with the Recon and fix that equipment. They can go fix fencing to be able to identify troubled fungi in the early season when the crops are low and wet, and you can't take a pickup truck or an ATV out in the dry season. When the crops are up to your shoulder to be able to get out to trouble areas can be troublesome. But then the secondary purpose is they don't only have to use this as a manned vehicle, they can actually apply for a heavy lift drone exception, a certificate of authorization from the FAA, and use this as a drone."

Like many other flying vehicle startups, Ryse Aero was short of the much-needed funding.

The Customers

As an added feature for the Ryse Recon, Kowitz and team created a 30-foot boom, which can be attached to the sides of the flying vehicles. Rather than a pilot, a large tank sits in the pilot seat, designed to carry fertilizer or other liquids that can be sprayed as the vehicle is remotely operated. Built-in technology allows certain parameters to be set, such as the vehicles cannot go past a certain point on a farm, or a certain height or speed. A farmer could fly it remotely, with 250 pounds of chemical in it. The PAV as a drone also could spray pesticides or herbicides, for example, with very specific nozzles. Smaller drones also could be used for spraying, but they typically can handle about 10 gallons of liquid with limited airtime due to battery constraints. Kowitz says the Recon as a drone can spray more than 12 acres an hour, depending upon what is being sprayed. However, the vehicle also can be used outside of framing applications.

"Our actual first buyer is someone who wants to fly," says Kowitz. "His office is one mile across a lake from where he lives, but he has to drive 25 minutes around. He just wants to get to his office in one or two minutes, so it saves him 50 minutes a day in order to do it. There are a lot of people who want to use it as part of their business to be able to look over their properties, sometimes for personal security. We're talking with the military as well for a number of manned and unmanned projects. There are a lot of different opportunities in a lot of different spaces, such as national park services, emergency response because you don't have to be a pilot to operate it. It really has some benefits as a lightweight vehicle.

"It certainly can be hobbyists and it still fits within the spirit of that. We have a fair amount of people who live on lakefronts and have large property who just wanted to tool around. I think even the farmer who buys it is going to use it for more than just surveying crops. He's going to use it primarily as part of his daily life, which is part of the spirit of the sport and recreational use of it.

"We've been developing aircraft in the air taxi space since 2015. Our team of engineers has been doing this probably longer than almost anyone. We probably were the first group to get a special airworthiness certificate long before Joby or anybody else. "We've been doing it for a long time. Our controls and the ability to understand how to build a system that other people can fly was probably a little bit shorter of a learning curve than a lot of these other companies are. And so being able to build this and quickly get it to market in less than three years has been a pretty amazing feat doing it within the ultralight space."

Test Flyers

Ryes Aero has been very market focused on the approach to its PAV. At Ryse Aero headquarters, there's a wall filled with pictures of individuals who have flown the Rysse Recon. Each person was photographed sitting in the vehicle before or after their individual flight. Before each flight, the soon-to-be-pilot receives a detailed briefing on all the details of how to fly the Recon and exactly what to expect and what do to in any potential emergency situation. The briefing is heavily focused on safety and safety procedures.

"You want to be as safe as possible, so we made sure we flew it a lot before anybody else," says Kowitz. "We have taken it to trade shows and we have let other people fly it, a lot of other people. We've had several reporters fly it and my wife has flown. A lot of people have flown it, for pretty good clips. It's kind of funny because I spend about 15 minutes teaching them and probably the bulk of that 15 minutes after you explained the controls, I have them sit in it. It's really talking about what to do if something goes wrong and emergency procedures like in aircraft, you want to make sure you know what to do in case something happens you don't expect, so that there is no unexpected. That's the idea of flying any aircraft. And we spend the majority of the time talking about if a motor goes out, if you get an audible message to X, if your screen starts flashing red, what do you do beyond that?"

Like most PAVs, operating the craft is very straightforward with most operations managed by joysticks. As part of the briefing, Kowitz has the pilot sit in the craft and they go over the controls and what the two joysticks do.

"It's really this is forward, this is backwards, this is side to side," says Kowitz. "And there's a thumb wheel that lets you spin around. The left stick is up and down. If you just let go, if you get scared at any time, take your hands off the stick. You're just going to hover just like any drone. It's going to stay right where you're at. Giving people that comfort level before they fly is fantastic."

Following the training, the pilot is bucked in with a helmet and visor with communications incorporated, so the pilot can speak with ground personnel for an entire test flight.

"After they get in the air and I'm on the phone with them talking to them, I walk them through," says Kowitz. "I say, start giving ease into the control. Get a feel for it. Okay. Give it a little more. Okay. Ease out of the controls. That's usually the conversation. I make them go through every direction and then when I'm done, I'm like, okay, what do you want to do? And they can go left and right, back and forth as hard as they want. It's not going to fall out of the sky. If they pull all the way back on the stick, it's just going to land nice and easy. If they try to go all the way forward, we've geo fenced it. They're not going up too high, so we really control where they're at with it. When they land after they've flown around for five or 10 minutes, I usually walk over and say, what do you think? And the response is usually 'I'm speechless.' That was so cool. And that's really the experience we want from people. But we want them to also walk away with, I can see a lot of use cases for this, not just flying me from my house into the city.

"We think flying it on your farm, flying it on your large property, flying it at your lake house, flying it at a ski resort, or an area outside of a national park is probably the bigger, better, and more constrained use case for this. It's going to be a conversation piece when people start to see these showing up in their areas. It's really not going to be neighborhoods as much as you're going to be driving along the highway and you'll see it zooming down along the highway on somebody's property. They're going to check out their irrigation tiles or they're going to check out the fence or maybe search for some lost cattle. That's where the use case is going to develop quickly. A lot of people see it. Being in the backyard, that's going to be further out. Even the air taxi guys are telling you that. I want to go fly one of these things. Yeah, well, come to my office. You're more than welcome to fly it."

Evolving the PAV

Kowitz understands the short- and long-range views of the electric aerial market and consciously decided to start in the segment that will be earliest, due to factors such as less regulation and ease of flying the vehicles.

"This is an entry point for us," says Kowitz. "We think we're going to be years ahead of everybody else with, with tons of vehicles in the market. Hundreds if not thousands of vehicles in the market because they're expensive right now. And the costs are going to drive down. So, we're going to have that personal vehicle experience as an ultralight fitting in the ultralight space pretty early on. But this is a platform, so using it as a drone, using it in military is a huge application that we see. But realistically, when you talk about the consumer market, 10 years from now the consumer market is going to develop significantly. Propulsion is going to grow; it's going to change. Your batteries are going to grow, they're going to change. The air taxi market is going to start off more like your helicopter charter service. There already are companies doing that right now, in New York City and all over the country. That's where air taxi is going to start over the next few years. It's going to expand as the costs come down and the National Airspace System (NAS) gets more developed.

There are several companies working with the FAA on that. It's not an easy lift. Where the air taxi market wants to go is they want to go to an unmanned aircraft. You and five of your friends hop in at Newark at Liberty and you fly into the city, and you land on top of the World Trade Center, and you hop out and you go to work. That's what they want with no operator in it at all. I don't think that's still 2040. I think a manned version of that in 2030 is going to happen. Are there going to be a lot of them? Not initially, but I think that market is a less attainable market than the personal market, the guy who can hop in it and fly it, pull it out of his garage, take off and fly 20 minutes to his office, not in New York City, but in the suburbs. That's really the bigger use case. That guy is going to want to be able to land it in his parking lot.

Just like we've started to see EV charging stations pop up in parking lots. I think you're going to see a pad pop up where you can come, you can land your vehicle, and then wheel it off to the side into a parking space so the next guy can come in and land his vehicle in your more suburban areas. That's really the bigger market and that's what we're playing into. We believe that's going to be the big market."

While both the personal aerial vehicles and the certified personal aerial vehicles are similar, there are major differences, as previously noted. When Ryse Aero started, Kowitz had a choice to go the PAV or the PAVC route, taking the former, providing the company with an earlier launch date for potential vehicle buyers. Kowitz has a realistic view of were both categories of aerial vehicles fit.

"They fit into the same market, but there are going to have to be some changes. For that single operator or two-person operator vehicle, you need a pilot's license to do that right now and I don't think a lot of people are going to do that. But I'm a proponent. I'm working with the flight school that we're trying to get the local high school involved and get more kids into aviation. And that's a huge factor in all this. But the guy who's got a full-time job, who's already making $150,000 or $250, 000 a year to take the time to do that flight, there's a limited number of people who want to do that. We do have to get to more of what the FAA terms SVO (simplified vehicle operator) market. And we're already there because we can already give it to you. But the FAA has got to get there when you talk about a heavier aircraft with multiple people hopping in it and flying over congested areas. That's a different game when you're a single occupant in an aircraft that you're responsible for and it's lightweight. And really, the only person you're going to injure is you. That's a risk you take on. And that's what FAA Part 103 was designed for. When you start getting into Part 91 (general operating and flight rules for civil aircraft) and Part 135 (non-scheduled air charter carrier) you're talking about heavier vehicles. That's riskier. It has to be locked down safe and that operator has to really know what they're doing. You're a pilot, so you understand restricted airspace and all the circles and everything on the, on your maps. The normal person who doesn't have a pilot's license may not be aware of that."

Flight Information

While the Ryse Recon is in the ultralight category of aerial vehicles, the technology involved does provide more than rudimentary flight information, which is the case with other powered lightweights. Part of this is due to the available technology, including sensors and communications, so that a large amount of information can be provided to a screen within the vehicle.

"Our system, for example, shows your airspace. We're a little more concentrated on our sale. When we go to a farm, that guy's typically using it on his farm and within a 10-mile radius. And so that 10-mile radius is pretty easy to say, this is here and you're in airspace here. Or you're not here. If you want to go there, you have to call air traffic control before you hop in it. It's easy when you start opening up and you're flying in cities and you're going to let people fly in multiple directions. If longer ranges, you're going to have to have a good national airspace system. For drones, we use the same technology and can say you're not allowed to fly there. Those PACV guys are going to have to have the ability to understand how to fly in that airspace. And again, I think there's going to be some new airspace rules, more visual flight rules quarters for people to fly through.

"This is a big lift for the FAA and they're going to want to do it right. They've done a great job in protecting us and protecting people who don't fly from getting hit. Rarely do you hear of an accident where people on the ground are injured, because we're all trained in how to work around that. If that level of training goes away, what's the vehicle doing? There are some complications that the FAA and everybody's got to work through together to get through. That's not going to be sleight of hand."

"We have geofencing capability, we geofence for perimeter, and we can geofence for altitude as well. When you're flying, when you're using our vehicle as a heavy lift drone, you have to lock that down. That's how you get the certificate of authorization. When we go around the country and do demo flights where we let you fly it, we put it in what we call bumper car mode, which is test drive mode, where you try to go up, you can go up, but at some point you can't go any higher. If you try to go forward, it's not going to let you go. You're going to have to turn it back around.

Market Position

Kowitz was comfortable with the Ryse Aero progress and market position at the moment. The company may be in the startup category, but it is quite far along that road. The Ryse Recon continues to be fine-tuned and improved with each new version on the way to the production stage.

"We're ahead of most people," says Kowitz. "We flew 45 times at the consumer electronics show this past year in front of more than 40, 000 people. We did not miss a flight. We flew every time, and we flew it for as much fun as you can. We flew down the Las Vegas Strip. We're the first aircraft to be allowed to fly down the Las Vegas Strip and the first aircraft to be able to fly at CES. And it's because the vehicle is so safe. Demonstrated flying. We've flown at farm shows all over the country and other events as well. We put a lot of time into it to get there. From that perspective, we're ahead of the market. What concerns me about the market is when people start talking about the use cases that are not realistic in today's terms, they're going to be years out. And we don't want to be associated with that. We've built a vehicle that we know we can fly today and fly safely today. But we put it in a box like you're not going 2,000 or 3,000 feet up. You're not flying at 130 miles an hour, 150 knots. It's a very controlled environment with a very specific use case, and we're training people very carefully on it."

Despite its progress along with the rapidly evolving PAV market, Kowitz has some worries about how a number of new PAV owners may be irresponsible in how they manage or operate the aerial vehicle. Despite the very clear rules and regulation set out by the FAA, there still is the potential for careless actions.

"My concern is people are going to start trying to do things without being in controlled areas, and they could ruin it for everybody. And that's, that's really the biggest concern I have is people who just come on the scene and think they're going to do this. Now, ultralights are designed for that, right? The novice can build their own aircraft and fly it and they're not hurting anybody. It's when people start talking about bigger use cases that aren't really viable today as if they are viable today. That concerns me. As a company, we're trying to protect our reputation, but we're also very much trying to protect the individual who's taking this thing out. We have a whole check off for documents before we release anything. There's a whole set of inspections every time that thing goes out the door, and we're very careful to follow procedures as if we were going through an airworthiness certification. I don't know what other people are doing, but our job is to protect anybody who goes in it that if they crash it, it's because they did something stupid, not because the vehicle didn't operate as expected. That's the biggest concern from my perspective, that you can do something stupid with this because you're mandated to fly it. It sees and avoids, just the same as a regular aircraft. But there's no technology in it that's going to look for power lines. Your eyeballs need to do it."

Liability

The FAA Part 103 rules for ultralights clearly state that the operator assumes full liability for operation of the aerial vehicle, which may not be commonly known. The seller or distributor assumes no liability and some of these vehicles are shipped for assembly by the owner, so this is an owner-beware situation. However, it is in the interest of the manufacturers and sellers of PAVs to ensure responsible training and operation for the sake of their business and safety of their customers.

"You're liable for this, as the owner operator of it," says Kowitz. "That's part of what we hold on to. I don't know if people generally know it, but we explain it. When you sign a purchase agreement, you're signing on and saying you understand the liability, and we are going to train you on all that. We're going to train you on airspace, on the emergency procedures, the safety procedures, inspection, maintenance as well as you listen. This is a deadly piece of equipment that you're operating. Yes, it's doing everything in its power to keep you safe, but it doesn't keep you from doing anything stupid. Look at ATVs. There are more ATV deaths annually than there are ultralight deaths annually, so it's all about training. Nobody's training heavily on an ATV, but ultralights, I think most people are getting heavy training on it. We want to continue with that. I think the demand is there, but when it comes down to it, flight is scary to a lot of people."

The hard market reality is that considerable funding is needed to get a program such as Ryse off the ground.

The PAV is the first category of flying vehicles to enter the market starting in 2025. The certified personal aerial vehicles, the PACV category, will follow in a longer timeframe due to required FAA certification, not expected any time before 2026, other than for test flights, which we discuss in the next chapter.

4 PAVC

The Certified Personal Aerial Vehicle

The certified personal aerial vehicles are in the development and testing phase, so these will be later to market than the PAVs, the earliest being in 2025. These aerial vehicles will have significantly more technological capabilities than PAVs and, in many ways, represent the very personal air travel future. The vision for these vehicles is that they can be stored in your garage like a car and for use would be rolled out of the garage and take off vertically from the driveway. This is the longer-range view, and lots has to be done before getting to that stage. However, that is the future now in full development mode. As in most of the early flying vehicles, costs are on the high side, no matter the category, with initial pricing in the PAVC category generally in the $300,000 range. As time goes by, expect more PAVCs to be developed, especially after final rules and regulations around the vehicles are agreed on and implemented.

As in all the flying vehicle categories, PAVCs will be launching around the world. Eventually, many of these will fly automatically, with no on-board pilot, picking up and dropping off passengers like a flying Uber, sans driver. This will be accepted in some countries way ahead of others, due to different regulatory environments and densely congested residential areas where road travel is challenging, which is discussed elsewhere in the book.

The Certifications

Unlike personal aerial vehicles in the ultralight category, as discussed in the last chapter, the process for certified personal aerial vehicles is totally different. The PAVC is essentially viewed as an airplane, which can carry passengers. This means the risk is not only to the pilot, as in the case with the single-person PAV, so regulators have a much more rigorous process before these aerial vehicles will be allowed to take flight. Additionally, airworthiness certification for PAVC is a rigorous process.

As part of any certification project for flying vehicles, the FAA states that it conducts the following:

- A review of any proposed designs and the methods that will be used to show that these designs and the overall aerial vehicle complies with FAA regulations.
- Ground tests and flight tests to demonstrate safe operations.
- An evaluation of the aerial vehicle's required maintenance and operational suitability for introduction of the airplane into service.
- Collaboration with other civil aviation authorities on their approval of the aircraft for import.

Coming Innovation

The current state of PAVs is anything but static, as innovation continues behind the scenes. Some of these innovations involved improving battery capabilities, efficiency in charging, and new fuels such as hydrogen. There also is innovation in the actual propulsion used in certified vehicles. One such technology involves replacing standard propellers with a different propulsion system.

CycloTech in Linz, the third largest city in Austria, developed an air car called CruiseUp by using cyclorotor technology in place of traditional propellers. The company is developing the propulsion system for a two-passenger aerial vehicle with the capability of going sideways and backwards, braking mid-flight, and hovering, a system in development for 10 years. Tapping that technology, Klissarov Design, a French design studio, created a single-person concept vehicle called the Acro, designed to be flown from large yachts. That vehicle, at least as conceived, could fall under the ultralight category if it is further developed. While this technology is still in the development and testing phase, it holds promise for yet another innovative approach to getting people into the air.

Speed in the Sky

Unlike the ultralight category of personal aerial vehicles, which have the FAA-imposed top speed limit of 63 miles per hour, certified PAVs have yet to have such limits. This has led to visions of high-speed aerial vehicles still designed to fly well below commercial aircraft.

Ace VTOL in Perth, Australia, was founded in 2022 and is developing a series of speedy flying vehicles. Ace developed an all-electric vectored thrust fluidics engine combining heat from a continuous arc into the central inductive airflow, allowing for increased thrust. The company is developing a wide range of vehicles, including the Trinity, which could transport people or cargo, and the GT, a limited-edition flying muscle car. The company is working on some of the most advanced technological approaches in the market. As is typical in the flying vehicle ecosystem, Ace also has various partnerships. It has one with vertiport developer Volatus Infrastructure for vertiport development in Australia (Volatus Infrastructure is featured in a later chapter). Ace also partnered with Aeroauto in Florida for sales of its flagship GT Slipstream, a two-seater EAV the company says can reach speeds up to 270 mph and fly up to 300 miles on a single charge. That vehicle is expected to be market ready by 2027. Whether the development of new propulsion systems or the implementation of high air speeds, the innovation for flying vehicles is a continually wide-ranging and moving target.

Long-Time Coming

While it may seem that when a large number of PAVCs come to market that they were just developed, since so many are likely to reach the market in the same time frames. However, PAVC companies have been working on developing their vehicles for years. A company typically started with an idea or a vision and laid some groundwork. Then a small-scale version of the idea might be created and tested to validate the initial approach, which could be dramatically changed based on test results. Eventually, a full-scale model is created and then that has to be tested, requiring the creation of remote-control software and systems to validate the system before any human-piloted testing starts. And there's the certification process, since the FAA does not want random test flights of vehicles it does not deem ready for testing.

A good example of how long it takes from start to testing is the AirCar Corp. in Istanbul, Turkey. The Klein Vision startup was founded in 2017 and started development of its two-seater flying vehicle with subscale models. The company partnered with Softtech to explore autonomous flight technologies. It then created computer simulation models working toward advanced machine vison systems. By 2020, the first full-scale model of the AirCar was created. It took from 2017 to 2023 to prepare the AirCar for its very first crewed testing while working toward the type certification process with the FAA and EASA. In March 2024, Klein Vision sold a geographically limited license to use the flying car technology to Hebei Jianxin Flying Car Technology in China. And in May 2024, company founder Stefan Klein took French composer and performer Jean-Michel Jarre for a flight in the vehicle.

Another example of the length of the development cycle is Air EV in Hefa, Israel, and maker of the Air One, also founded in 2017. The company's Air One flying vehicle was introduced in 2021, conducted hover tests in 2022, and by mid-2024 had pre-orders for 1,000 of the $150,000 EAV. The fixed-wing, two-seater has eight motors, can fit onto a trailer, can travel at speeds up to 155 miles per hour, and has a one-hour charge time. With the regulatory requirements for certification, the vehicle is not expected to be available to the public until 2026. In late 2023, Air EV, which bills itself as Air, signed a deal to join the U.S. Air Force AFWERX Agility Prime program with the Air Force considering using the Air One for logistics missions. That process started in 2022, highlighting the time it takes to advance aerial mobility innovations. The startup flew its two-passenger Air One flying vehicle in Florida at a private demo day in late 2025. As in the development of most of the certified personal aerial vehicles coming, the overall process takes years before hitting practical reality, the price it takes for getting it right.

Creating a PAVC

Developing personal aerial vehicles that can pass all the required FAA certifications can take time. A lot of time. Prototypes are created for testing, generally first flown remotely to validate flying capability, and evolved from what is learned in early testing. These early iterations can be crude, bare-bones models just to validate various capabilities. Early tests can include tethering the vehicles to the ground, just in case. What was learned can cause a company to totally pivot in a different direction, which could be unexpected at the start. Company executives I spoke to during research of this book often said the process took much longer than they initially expected. Developing a PAVC is much harder than it may look from the outside. Executive of the vision or idea also takes a team of people, including engineers, software developers, and pilots. Doroni Aerospace in Pompano Beach, Florida, a beach town north of Fort Lauderdale, has been working on its certified personal aerial vehicle since 2016. Doron Merdinger the founder and CEO of Doroni Aerospace who had the initial flying vehicle vision long ago.

"It was a moment of inspiration, and a flying car was always my dream," says Merdinger. "I'm sure it's everybody's dream. I used to build and create in my room, because I wasn't much of an athlete, so I had to choose a different way to occupy my time. My world was always creating. I was always building stuff. Seven years ago, I saw this kid play with a drone when I was stuck in a traffic jam. I was like, why can't we scale it up? And this is how we hit it off."

The PAVC Vision

Leaders of flying vehicle companies typically had a burning vision of creating a flying vehicle of one sort or another. As you can see from numerous examples in this book, there is not one-size-fits all approach to flying vehicle development. Some take the ultralight approach, some go for vehicles that can fly and drive on roads, and others look at passenger-carrying commercial vehicles. No approaches can be defined as wrong; they just are different. No matter the approach, every flying vehicle company leader I spoke with had a solid vision of what their flying vehicle would basically look like and effectively do. At Doroni Aerospace, the vision of the eventual popularization of the personal flying vehicle of the future is a driving force. Starting by knowing everything about creating aerial vehicles was hardly the prerequisite to get underway.

"I didn't know too much about aerospace, but I knew a lot about materials, technologies, and about fabrication process," says Merdinger. "On the design, I started researching and started going online and understanding who my team would need to be. The beginning was about understanding the propulsion system and how it was going to work. Are we going to go forward with propellers with electric motors? What are the options? I found out pretty early that the technology was very premature at that time, in terms of batteries and motors, so I had to handpick those parts. But before that I had to find like-minded people, kind of skunkworks people. I was able to find them in the United States and outside. Then we started defining what we were looking to do. How is it going to shape up? What's the flight control system? Things were coming together. I like to pose a question and then challenge myself to answer it. I'm playing devil's advocate with myself and if I don't answer and if it's not good enough, I challenge people to challenge me and I'm playing all the time back and forth until I think I'm getting this feeling that this is it and you can't milk it anymore."

Personal Testing

Once Doroni Aerospace had a prototype that could fly, Merdinger wanted to try it out himself to feel the experience. At that point, the vehicle was not yet certified to be tested outdoors. Merdinger decided to test fly it indoors at their facility. Flying or hovering indoors can be even more challenging than outdoors, since all the wind currents from the propellers stay and swirl inside the building. The company figured that if the vehicle could fly stably inside the building, it would then work outdoors.

"It was very important for me to get the feeling of flying it," says Merdinger. "I knew when you create things, at least for me, I already have it in my mind. I kind of feel it. But I had to actually see and feel how it flies. I sat inside and I knew the systems because I was part of the team. And I trust my team, so I just went inside and flew. It was like sitting on a cloud and very, very comfortable, and it felt very safe for me. Because when I was in the Israeli Air Force, one time I was flying in a helicopter and I didn't like that feeling, with the vibration. There were a lot of opposite forces here in our test, but everything was much smoother. Also, because it's electric, it's different. The experience was second to none and I can't wait to get the final product out there. My biggest frustration is that I want to present it to the world."

Redundant systems are in the DNA of electric flying vehicle developers. Since the vehicles generally have multiple motors, multiple propellers, and multiple batteries, designers tend to keep the systems somewhat separate, avoiding any single point of failure. Vehicles typically can stay in the air or safely land if a particular system fails. Doroni Aerospace was no exception.

"We did a lot of redundancies and a lot of ergonomics and put a lot of thought in the process," says Merdinger. "The idea was that it will be very intuitive, very simple to use, and will reduce pilot error as much as possible, even given that it has an autopilot, which serves to keep it in the air with a gyroscope and multiple of other sensors."

Parked in the Garage

In the future, Merdinger expects large individual ownership of vehicles such as his. "Just go outside and see how the kids are flying drones today. This is how it's going to shape up. And the auto pilots take control of everything else. This is how we look at it, as it's very intuitive. The idea is that you will be able to park it in a two-car garage, and you charge it there. Then you can tow the vehicle out backwards a couple of yards and then just go vertical. Very simple. This is something that we're working on as well. It's just almost self-sustained. You would know you're at the altitude that you want to reach, just taking off, just give you a full clearance. And you check all the systems, the system will run diagnostic by itself, you do a visual as well and then you just go up. There are a couple of ways to fly it. One of them is with a stick, the other one is just putting in the address, which is how we're developing it as well."

Another part of the Doroni vision is that flying vehicles are about experiencing the environment and not just traveling between two locations. PAVs and PAVCs are not intended to be flown at high altitudes, for both regulatory and technical reasons. "Like a drone, the higher you go, the more energy you need to use, the time it takes, and the air becomes thinner at a higher altitude," says Merdinger. "The experiences for me is I want to be still engaged looking at the surroundings like a drone. Unless you really need to, you shouldn't go there because the whole idea is really the experience more than about saving time. You want to see everything around you and see how beautiful our Earth is."

Pre-Orders Galore

Like other flying vehicles makers, Doroni has numerous pre-orders for their vehicles once ready for delivery. "We have more than 320 pre-orders or requests," says Merdinger. "People are putting money down and we put it aside for when we start delivering those units. It's exciting how people are enthusiastic. I really want to make it affordable. This is my goal to give democratize flight, to make it more affordable, then really depends on the system with the beginning. We wanted a different target price and we had to go up with the costs changed because of the global pandemic. "We're making like the most advanced vehicle, but it's not that advanced because it's still using electric motors and propellers. It's not like we're doing anything like electromagnetism, but we're doing something which is wonderful and sustainable in a very big way.

"We all know the battery technology keeps improving. We started with the first iPhone, it was 30 minutes and now it's like days. The same with cars. In the beginning, it was doing 50 miles, maybe even less, and now it's hundreds of miles. We know that the better technology is getting there, the electric motors are getting there. This is the first thing and people are more aware of what's going on, in ecological system, the world is warming up. Why do we need more roads? Why do we need to continue to create asphalt and concrete and destroy our habitat? This is just something that I don't understand. The United States is a little bit far behind. South America even is more advanced, as well as Brazil, Australia, and Europe. And definitely China. Now we are trying to catch up. There's a big push with the federal and local government for it, for all electric and also the FAA is working on how to facilitate all the vehicles in the air co working together. It's a big thing and things are happening and it's not just one company. People ask me, why didn't it happen before. If the technology was sustainable 20 years ago, there was not much public acceptance as today. This can work today because of many exterior systems, like geofencing, kind of invisible walls, which cannot even pass. It's easier even controlling irregular aircraft from the FAA perspective. They just created some sort of invisible wall, so you cannot even cross areas that are important, like the airport or federal government or army bases."

Certified for Flight

The details are still to be determined in what it would take to fly vehicles such as Doroni's. It could take a driver's license and medical clearance along with a course of at least 20 hours, with 15 hours of flight experience. Or it may take having a private pilot's license. In the case of various fling vehicle makers, it is not pilots that dominate the pre-ordering activities. For the Doroni vehicles, Merdinger says they have doctors as well as people who want to fly it to the golf course. People also could potentially fly from land to their yacht, as with the case of the Klissarov Design concept vehicle Acro discussed earlier in this chapter. The vehicle could potentially be used as a joyride in an amusement park or to check high electric wires. The goal for Doroni Aerospace is to be in production mode in two years. The FAA has certified the Doroni vehicle for official testing, so that process is underway as Doron Merdinger's vision for his version of the flying car takes shape. In the next chapter, we discuss the actual flying cars in the works.

In 2025, Doroni Aerospace received a $30 million strategic investment from Kingdom Aero Industries in Saudi Arabia.

5 Flying Cars

The New Air Road Travel

Some developers of flying vehicles refer to their crafts as flying cars, even though they cannot drive on roads. This is because some view their flying vehicle as the car of the future, being wheeled out of the garage and taking off straight up from the driveway or the front yard. After a quick trip, the vehicle would be driven (flown) home and get parked back in the garage, just like the car of today. That is the long-range future of flying vehicles, though realistically most are a number of years to come.

While flying vehicles sometimes are referred to as flying cars, there actually are flying cars being created. These are passenger-carrying vehicles that can be driven on the road and then self-convert to become flying machines. These are already doing street and flight testing. Like other flying vehicles being developed, the first versions of flying cars are just that: functioning prototypes to test various aspects of the vehicles, from road performance to aerial takeoff and maneuverability. Some of the first models are large and tend to cause people to stare when they see one drive by. When on display in various showrooms, they easily attract curious onlookers. However, this does not mean that what is seen is the final product, since much fine-tuning is happening to all certified flying vehicles before the production models are out.

Cars that Fly

To somewhat confuse the market even more, there are two different categories of flying cars. There are those that comprise one complete vehicle that can drive on roads and have the capability to take off and fly. In these, the wings could fold in or rest on the top of the vehicle for road travel. On the road, most of these are not be fast as a typical car since they are carrying all the necessary equipment for flying. The shorter-range approach of these is to validate various elements, such as road worthiness, speed, dealing with traffic, weight, fuel sources, and power sources. For aerial testing, components could include road elements while in the air, distance capability, speed, takeoff and landing requirements, and overall weight. The obvious first impression of a car that can also fly might be an expectation of a compromised vehicle that does neither very well. What may come to mind are duck boats, those vehicles that can drive on roads and operate in water, typically used in city tours around the United States. The duck boats meander along city streets not as nimble as cars, motorcycles, or even trucks, but quite similar to a large bus rather than a car. Duck boats also have a propeller for powering the vehicle in the water. The vehicle typically enters via an easy launch ramp, navigates at about idle speed around a body of relatively calm water, and then exits back onto the road, converting back to street operating mode. Duck boats are neither fast nor nimble, but they get the job done. Expectations are that the duck boat is not great on land and not great in the water, but neither needs to be great to fulfill their tour missions.

Flying Car Expectations

Developers of flying cars are fully aware of the potential of creating vehicles that are perceived to be neither great at road nor air travel. As a result, they aim to create vehicles that go well beyond the expectations of being marginal at both ends. For example, Samson Sky, maker of the Switchblade flying car, set out to counter this idea that a flying car is going to be a mediocre car or a mediocre flying vehicle, as detailed later in this chapter. The company's flying sports car can drive on land at speeds up to 125 miles per hour and fly at 200 miles per hour in the air, hardly a compromised flying car approach.

Cars that Are Flown

The other approach to flying cars is modular. For these, the road vehicle is a separate component from the flying part. One way to look at it is that there are road vehicles that fly and there are road vehicles that are flown. A good example of this is a prototype vehicle introduced in late 2023 by XPeng Aeroht in Guangzhou, China, a subsidiary of the automaker XPeng. In 2025, the flying car maker changed its name to Aridge. In its modular version, the five-passenger road vehicle has three axles and six wheels. When the large mini-bus type road vehicle stops, the flying portion comes out the back, separating itself from the road version to fly. Aridge also introduced a land aircraft carrier that switches between aerial and ground. Both the modular approach and combined land-air vehicle are in full development mode by various startups with no final determination of the post-prototype vehicles.

Double Approvals

The obvious challenge for flying cars is that double certifications are needed. The FAA must validate the air worthiness of the vehicle to carry passengers, a process not expected until 2026. However, like other certified flying vehicles, flying cars can receive certification for flight testing. While 2026 will be a major year for personal aerial vehicles -- the powered ultralights that are coming to market as discussed earlier in the book -- it also will be the timeframe for prototypes and testing of all types of certified vehicles, including flying cars. In addition to FAA certifications, flying cars also need approval from the local vehicle registration departments for the vehicle to be driven on roads. In the case of flying cars, sometimes these may technically be registered as three-wheel motorcycles, bypassing traditional automotive requirements. For example, the Pal-V flying car Liberty is a three-wheeled, fixed-wing vehicle that turns into a gyroplane. A propeller at the back pushes the two-passenger Liberty forward, making a rotor on top of the vehicle spin. The $500,000 Liberty has been in road testing since 2020. In the true global nature of the electric aerial revolution, Dubai company Aviterra in 2024 ordered 100 flying vehicles from Pal-V to bring the flying vehicles to the Middle East and Africa.

Time for Testing

There are several flying cars being developed and tested, a process in some cases going on for several years already. For example, Alef Aeronautics in San Mateo, California, started in 2015 with a drawing on a napkin in a café, with a vision of developing the car in about six months. That turned out to be eight years. This is another example of the complexities of developing a validated flying vehicle in any category. While the Alef Model A is not planned for delivery until 2026, the vehicle was listed by Time as one of the best inventions of 2023 and a prototype was featured at the Detroit Auto Show in 2023. Like a car, the Alef has four tires for driving, and for flying, and eight independent propellers. The $300,000 vehicle is expected to be able to travel distances of 200 miles on the road and 110 miles in flight. Alef has received special air worthiness certification from the FAA to conduct vehicle testing.

Deliveries Down the Road

Flying cars will continue to be developed and evolve as they go. As is the case with other personal aerial vehicles, flying car companies typically take preorders. While deposits are held in escrow, the number and scope of preorders gives a flying vehicle startup a benchmark for market adoption of its coming vehicles. Aska in Mountain View, California, was founded in 2018, creating its first concept design vehicle a year later. By 2021, Aska started taking preorders for the $789,000 Aska A5 flying car and by and started the FAA Type Certification process a year later. A fully functional prototype was featured at CES in Las Vegas in 2023. The A5 has been driving around California roads for testing and also has received FAA special airworthiness certification for flight testing. The wings and propellers of the electric vehicle fold in and rest atop the vehicle for road driving and the prototype size is like a large SUV. With six motor systems, the A5 is designed to fly 250 miles at 150 miles per hour. Even though preorders started in 2021, deliveries are not expected to start until 2026, yet another indication of the amount of time necessary to get a flying car approved and into the market.

The Fast-Flying Car

Developers of flying cars must counter the perception that vehicles that drive and fly require a compromise on one or both aspects. One company had this realization when it started and aimed at exceeding expectations on both fronts. The Switchblade flying car, a hybrid vehicle by Samson Sky in Redmond, Oregon, is the result of all that focus. Sam Bousfield, Founder and CEO of Samson Sky, has worked on the creation and development of the Switchblade for the past 17 years.

"I had just gotten done with a bit of work with some Boeing engineers on a high-speed aircraft design and then I was looking at the future of things," says Bousfield. "I started looking at what the future of transportation is going to be. What I saw is the future as portrayed in Star Wars, Star Trek, The Fifth Element, that kind of thing, but pretty much all in the air with the ground reserved for walking or the occasional delivery of goods. But I saw that as a future. Even in the Jetsons, most of the Jetsons, when they first put that out, they had escalators, they had phone watches, they had all kinds of things that all have come to pass, except for the flying car. So, I said, we have to do that, so that's what I looked at. The future of travel is in the air. My thought process took me quite some time and figured we're going to need something that does a little of both, because you still have so much happening on the ground, and you've got to interact with that, and in the air, that's the future. It's something that would do a little of both."

Staying the Course

Bousfield has a background in architecture and aeronautical design and is the author of numerous patents, so he's comfortable with inventing things. However, as is the case with many people we spoke with during the research for this book, developing a flying vehicle can take much longer than initially thought. Bousfield says if he knew it would take this long, he may not have started. But once he starts something, Bousfield tends to see it through to the end.

"It was a bit of an awakening to me to see how hard it was to actually do something that had to be good at the ground and good in the air," says Bousfield. "One of our biggest hurdles was to counter this idea that a flying car means that it's going to be a mediocre car or a mediocre plane or both. We didn't buy that, and we just didn't agree with it. As a result, we set the bar for high performance in both modes. It took a little longer than we thought to get there, but we recently flew and there you go."

The Switchblade has one engine that powers the wheels on the ground and a propeller for power in the air. The power in the wheels is used for takeoff, since the wheels can make the vehicle accelerate much faster than the propeller. After the wheels are used to get up to takeoff speed, they are phased out and the flying car takes off under propeller power. It lands using the same processes and does not need much runway. But unlike some of the other flying vehicles that take off and land vertically, as discussed earlier, this flying car is designed to take off and land on runways, such as at regional airports.

"We can land in 700 feet and take off in about 1,100 feet," says Bousfield. "It's taking off a little faster than that in our test flights, so we might be better than we thought. There are 5,000 public and 8,000 private airports around the United States and most of the public airports are at least 2,000 feet long."

Evolving Market

Bousfield says his company has received more than 2,700 reservations from people in more than 55 countries, including at least one person from each of the United States. The orders come from people who heard about the Switchblade in one way or another. Samsom Sky has conducted its marketing only via Facebook advertising, but the company has received widespread media coverage. The target market for the Switchblade also has evolved over time.

"It was initially aimed at pilots, because originally, we thought pilots were our market, says Bousfield. "There are 600,000 pilots in the United States, which is a fair-sized market. We could live with that. That's what we found in surveys when we first started and just surveyed existing pilots and found a good response from them so that's really where we were heading. We did find over the last few years that more and more non pilots were reserving Switchblades and it kind of floored us, because we didn't know how they're learning about it. That caused us to take a new look. We did another round of surveys about a year ago and we did the existing products again, just to have a side check. We also researched and did market surveys into this group of 21 million people who are regional travelers in the United States. That's a big number and they take a lot of trips, like 189 million trips a year or something like that. The business or personal trips are in the 200- to 500-mile range. A lot of them are just people visiting relatives or going someplace, so there's a spouse or a special person.

"I think it would be an awesome business tool. If you just look at it, if you're in a space where you've got another big city, but it's like 100 miles so it's hard to reach. But suddenly, you have something that will get you there in half an hour. That could make a good case for it. Or guys who usually spend their life on the road, such as sales guys or service techs, where they're constantly a week away and then weekends at home with their family. This could make a big difference for them. Because now they can go out, make two or three calls, and then still return at the end of the day to be with family and be home almost every day. That would go a long way towards enhancing people's lifestyles."

Licensed to Fly

When Bousfield started his flying car project, he was not a licensed pilot. Along the way, he decided he also wanted to eventually fly a Switchblade, so he attended training and received his private pilot's license.

"I wasn't a pilot when I started it," says Bousfield. "Funny thing I learned as I went in, and probably three years into it I realized I'm designing an aircraft, and I don't fly. I think I can do something about that. So, I started learning to fly, and I've been flying since and really enjoy a lot of flying. That took about 40 hours of in-the-air flight training and some ground school. But the nice part about the Switchblade is that you can buy the switchblade and use it as your aircraft for flight training. It's smart because a person's first plane they fly and train in their first time sinks in because they're absorbing it all. As they are getting into flying, they remember their first plane. Often, you'll save about half the cost of getting a license."

Bousfield says several companies want to send their flight instructors to Samson Sky to get checked out on the Switchblade so they can help people learn to fly in it. For road travel, the vehicles can pull up to a local gas station to fuel. However, the vehicle had to be modified for which fuel would work in it. By June 2024, the Switchblade already was authorized to drive to airports on state roads and highways in New Hampshire and Minnesota, with other states poised to follow.

"Most aircraft still use leaded fuel," says Bousfield. "It's low lead, but it still has lead in it. We could not imagine that the EPA (Environmental Protection Agency) would be okay with us driving around neighborhoods burning leaded fuel. We had to come up with an engine that would be suitable for the air, but also work for the EPA and be reasonably responsible for things on the ground. It took a little while and we went through three engines and finally got one and had to modify it. We modified it with two computers, two fuel pumps, two water pumps, and a very special tune on the on-board computer for this particular use. We took over the engine even though it's made by another manufacturer. We take over the warranty and the liability of that.

"Even though it's an experimental aircraft, and not a certified through the FAA, we still designed and tested it to certify standards. We built some of our own test equipment to do that. My personal favorite was a drop test for suspension. It is the most grueling test. You swear you're going to smash it to smithereens. You raise it up on a guillotine like thing with a ton of weight on it, to match what the FAA requirements are for our speed and weight. Then you just pull the cord, and you cross your fingers on the way down, it doesn't get smashed on the ground. After everything's cleared, after smoking, it was like 'holy cow! We passed everything the first time. That was really, really good. I don't know anyone who has ever done that, so hats off to us.

Challenging Production

"Production is the toughest nut to crack and that's what we're pushing on right now. We've been wrapping up the flight testing to get it up and get that final flight video that we could show the world. We flew and it flies just fine. The flight looked very normal, and the test pilot said the same. He says it's an honest aircraft. You tell it what to do, and it does it, so I'm happy with the results. That doesn't mean we aren't going to improve a few things. We had to show that it flies and improve as we went along all these years. One of the advantages of having lasted so long and taking us a bit of time to get this finally done is that technology has increased quite a bit in 15 years. There have been some improvements in avionics and the flight instruments. We even have a parachute as one of the things you can get as an option. But it weighs 55 pounds, so it takes a lot of fuel. If you want to have the safety of extra fuel, you just forego the parachute. In avionics these days, one of them has a little button you can push. If you run out of fuel in the air and your engine quits. You push the button all this time you're flying along and it's looking and seeing what's around you airport-wise where the wind is coming from, and how far you are off the ground. It then says 'okay, based on what I see here, this over here is the right airport for us,' and it starts gliding you right to that airport to land and radios ahead and says 'hey, we're having troubles make way, here we come. 'Meanwhile, that pilot can sit there and say, 'what happened here and why the engine quit.' That is probably a more successful way to go, and I think almost as good as a parachute.

"The next thing is building three production prototypes. And then try to break them. We'll tour them out in the countryside and give people rides for test drives and promote them. You want to break things early. If we're going to be setting up production and have 100 of them and they still have that one piece that is going to break, then you must replace it or you have unhappy customers."

The initial plan for Samson Sky was to design the Switchblade as a hybrid electric vehicle and eventually switch to all electric. When the company started, the motors available were not powerful enough to deliver what was conceived for the Switchblade.

"The batteries still aren't good enough to handle the kind of performance we have as a mission vehicle," says Bousfield. "We are putting out a hybrid electric version for production. The first production units will be a hybrid electric, and we'll go full electric as soon as we can. We'd like it to be just the batteries, but they are not yet strong enough or light enough to count for much right now. That's where we're headed as far as production goes."

Bousfield expects the Switchblade to be available by 2026, following production engineering and many flight tests. The plan is to then start with a limited-run production, to meet the requests of the people who pre-ordered. Working on improving the flying vehicle is set for all of 2024, the same timeframe as numerous other flying vehicle developers. After a 15-year flying car development journey, Samson Sky is hardly about to rush a product to market until totally ready.

Flown Cars

While the Switchblade by Samson Sky is a flying car, there also are vehicles that take to the air carried by a flying mechanism, such as the modular vehicle from XPeng Aeroht in China, discussed earlier in this chapter. There also are efforts in the United States to create a detachable flying car. LuftCar in Orlando, Florida, is developing a hydrogen-powered eVTOL flying car with a modular design. An advantage of LuftCar, which can carry up to nine passengers, is a longer cruise distance capability with a range of 300 miles. The flying mechanism referred to as a flying forklift connects and lifts the vehicle to be flown. The modular design allows the LuftCar road vehicle to detach from the airframe after a flight. The appearance in the air would be that of a plane that has picked up a car and is carrying it as the wings and propeller sit atop it. The company was at the seed funding stage in late 2023 as it focuses on creating a scaled prototype. Santh Sathya is the CEO and Founder of LuftCar.

"It's a hydrogen powered electric, vertical takeoff and landing vehicle," says Sathya. "It's more than a traditional eVTOL if there's anything traditional about eVTOLs. It's a unique airframe that is capable of carrying cargo, which could be a life cargo or just containers. These are cars, these are big trucks, there will be buses, cars, and this will be an ambulance that you can airlift. Our airframe behaves more like a flying forklift. What we're basically doing is building a drone capable of carrying road cars. The road cars can go from point B to point C to do the door-to-door delivery services."

Automated Flying

The LuftCar approach is to create a self-flying carrier that delivers a road vehicle to a location from which it then could be driven. The vehicle carried only would have to be adapted to the LuftCar flying forklift system, meaning potential vehicles from outside of LuftCar potentially could be airlifted.

"The vehicle is going to be autonomous," says Sathya. "We are putting together a cutting-edge AI autonomous package. It would include depth perception, distance perception, collision avoidance, and sensing of vehicles around as well as knowing emergency landing positions. We are putting together a solid AI architecture for autonomous piloting. In many ways, this is the future version of a flying car. It's a flying thing that picks up a vehicle or car or truck. It could be any vehicle but with some amount of customization on the subframe. The vehicle will have a cage-like design around the subframe, so that it can be lifted from the top and the bottom. We lift the vehicle with the five-point fold mechanism so it's a top hold. The bottom fold is very much like a pitcher carrying a baseball. The vehicle will be carried from point A to point B and any vehicle can be customized very much like how you add a trailer tow package to an existing truck or car. That's how this customization would be done."

The Inspiration

"The real inspiration was many, many years ago, when I saw the James Bond movie Man with the Golden Gun," says Sathya. "I was very awestruck by that idea of mounting air wings to a car. What they really did in the movie was they actually took an airplane and made it look like a car. But in our case, we cannot be thinking about how to really make a car fly.

There have been many companies that have been trained to airlift cars, and they have been lifting cars from the top. When you do that, it becomes a logistics nightmare, because of landing, it adds more complexity. Baseball is an inspiration, so I looked at the baseball hold. I looked at how Eagles carry their prey. They have a very natural way of lifting things that move and that's the kind of design that we are simulating in this. It's an inspiration from nature and inspiration from sports."

LuftCar initially considered the flying car approach but ultimately decided the separate vehicle design was the way to go, primarily due to potential inefficiencies the other way. The LuftCar way essentially looks at the flying component and the road vehicle as separable.

"The retracting wings design is something that many people have tried as well and we could easily do that too," says Sathya. "But we are trying to stay away from that model because we believe in energy efficiency, traveling efficiency, and logistical convenience. Taking the wings with you on a road environment is like putting more additional load on the powertrain and you're asking for a burning motor. On one hand, you're trying to use battery and minimize energy consumption. On the other hand, you're trying to take more payload as you drive on the streets. So, we have lots of regional airports, we can easily attach wings to these cars and then attach and drive off and let somebody else come and rent the wings and go. We are building an interchangeable road module, which in turn, enables customers to keep renting the wings as they go by. There will be new way to think about transportation. This would work and, logistically and energy wise, it is more efficient than the retracting wings design.

"This could be about transporting road vehicles or in the military case, you can transport tactical, small logistics vehicles across contested zones from one point to another, and safe land them in bases or take air ambulances to the frontlines. I can take doctors to the frontlines, but it could also be taking people and cargo from one place to another and then delivering cargo from that place to more of a door-to-door delivery."

LuftCar is initially targeting airports to start operations, a common approach by the coming air taxi services, as discussed later. Airports already have fully functioning operation facilities, including air traffic control, security, and emergency services. Airport runaways are less needed for aerial vehicles, since most take off and land vertically.

"We are targeting airports to start, because airports give you the infrastructure space to put hydrogen refueling," says Sathya. "That's the only reason why we want to have them land and take off in airports These vehicles also serve people for delivery, so we want them to land on reasonably flat surfaces so there is still an alignment between our flying and road module. It could be in a cul-de-sac, and we'd be in designated parking structures where we could build small pads where these vehicles can land, and they can serve the community. They can land anywhere, but most likely refueling will happen in designated locations, which would include regional airports."

The electric aerial revolution is about so much more than personal flying vehicles and air taxis. It involves a coming transformation in logistics and supply chains, with changes in warehousing and distribution needs. Delivery effectiveness can be improved with modifications in delivery times and locations. For example, MightyFly in California created a fully autonomous eVTOL that can automatically open and close its cargo bay door, load and unload cargo, and position packages totaling 100 pounds without human intervention. The vehicle then can fly itself up to 600 miles for package deliveries. These types of autonomous aerial operations will have a profound transformation on logistics and supply chains. The LuftCar approach highlights some of the coming shifts.

"Our primary use would be in two places, the first being cargo," says Sathya. "In cargo, it would be a company like UPS, FedEx, or Amazon buying a fleet of our vehicles, which can cover 300-mile radius cargo deliveries. They would not need as many warehouses as they do currently, and they could serve remote locations because they can fly their road cargo vehicle from the warehouse to a destination 150 miles away. The road vehicle can detach and then deliver. We calculate the cost would be one-tenth of the current overnight delivery for UPS or FedEx. The second use case would be in healthcare. We do have a letter of interest from AdventHealth for air ambulances. We can take ambulances to serve immediate, urgent care. People that are in secondary care can also still use these vehicles to come to the hospital so the doctors can see them. We see plenty of use in the military, particularly flying these vehicles from ship to shore and shore to outpost. Typically, you will carry road vehicles to some base and then from the base, drive the road vehicles from there. In this case, this will be just a one-wing system that can move road vehicles from a ship to a base and from the base, you can drive the road vehicles. There are lots of logistics applications in the Air Force and Navy and we are talking to Air Force and Navy now.

"The key differentiator here is hydrogen because hydrogen gives you the ability to carry heavy payload through longer distances. Hydrogen fuel cells typically have four times the energy density at the systems level compared to batteries. Because of hydrogen, you can achieve 300-mile ranges. That's the baseline range you're targeting, and it can go up to 500 miles as well. Then you keep refueling with hydrogen. The goal is to have hydrogen infrastructure along the way, and that hydrogen ecosystem is developing as we speak. Vehicles like ours provide a good offtake for those hydrogen infrastructure projects that are coming up across the globe. The cost of hydrogen keeps going down. And hydrogen is more of a net zero emissions in total compared to batteries. We think every eVTOL company that is serious about regional transportation will eventually have to look for the hydrogen option, unless some battery chemistry shows up that competes truly with hydrogen energy densities."

LuftCar is working to create protypes of its aerial vehicle and continue to follow the new aerial vehicle market. Since moving things comes with less risk than moving people, the startup is keeping its focus there. Moving cargo and serving first responder needs are going to be the early use cases for many certified aerial vehicles globally. This is already happening from countries outside the United States, such as Jump Aero, which has created eVTOL vehicles to transport medical personnel and equipment as a first responder service.

"We are right now building our prototype in Florida, and another prototype that we'll be building in California," says Sathya. "Between those two prototypes, we have committed to the FAA that we will send them for certification and validation. We plan to have these vehicles serving the market from 2027 onwards. Our hope is that we will have fully functional prototypes available for FAA certification from 2025. There are other global markets that we are looking at as well. There are different certification regulations across the globe. In some markets, we can get certification sooner. One of my personal goals is to have this vehicle serve our country first. We plan to serve the military, and then serve other markets and then expand further. We're targeting a 2026 to 2027 timeframe.

"Based on my experience working in different, large corporations, I want to listen to the market. I sensed the market two years ago when we started the company, and since we set off in the direction our vehicle has opened many different use cases. Hydrogen is opening up more opportunities than originally thought. It's opening a big global amount of opportunity for us, so we are going to listen to the market. We already have a purchase order for 15 vehicles for air cargo. We will be leading the customers to help us build these vehicles in the right numbers. And that's how we plan to scale up.

"We are initially planning for the cargo market and as our certification and the state of autonomy mature. Also, with the ability to override autonomy and ensure passenger safety and aligning emergency conditions and that that system is all being validated. We will start putting people on it and approach the regional transportation market. But our initial market should be cargo."

While the flying forklift is planned to be self-driving, when the road delivery is made, a driver may not be needed there either. This is because autonomous or self-driving technologies have been advancing for years, as I documented in my last book, Digital Transformation 3.0. As those capabilities continue to improve, LuftCar plans to incorporate them into its road vehicles.

"The road vehicle will also be autonomous to the extent that it can be autonomous," says Sathya. "The whole docking mechanism is going to be autonomous so the vehicle that lands can detach and go to the destination. I drive a Tesla to level three autonomy right now, so if I have to have a LuftCar, we'll be having a level three autonomy. Hopefully, in two or three years from now, we'll probably have level four autonomy. There will always be a human at first for the next

10 years to override autonomy to whatever extent, but it will be a combination of human and machine.

"It could also have a remote flight operator for different applications for gravity for flight operator, but for cargo, maybe does not have a person, depending on the state of autonomy. That's what I mean by we will listen to the market and the technologies that are relevant. We are not trying to develop our own autonomy from scratch. We are partnering with companies that do autonomous solutions, and we're integrating them into our vehicle.

"The market is exploding right now, in terms of the manufacturers proliferated out there. This is very similar to the 1950s in the automotive industry, with a lot of car makers putting motors on the buggies. Then they all get assimilated at one point. The market dictates who survives and who does not. It's not always based on how much money you raise. It all is based on how you resonate with the market and how your specific technology really fits the market needs or creates disruption in the market. We believe that we are very strong, and you need technology. We will continue to be able to converse on time and we hope we will have our own narrative. We are not psyched out by all the big players that have raised lots of money at this point in time, and a lot of small players too.

Demonstrating

LuftCar plans to use its prototype as a demonstration vehicle. The company is also creating a two-seater vehicle with plans to apply for FAA certification for that model as well. Additionally, in early 2024, LuftCar formed a strategic partnership with eFrancisco Motor Corporation (eFMC), a major jeepney maker, to develop flying cars for the Philippines.

But the core focus remains hydrogen and its future as a vehicle driver.

"We are participating in some very good government grant programs on the hydrogen infrastructure that as you build the vehicles, we are also building a refueling infrastructure," says Sathya. "The Holy Grail in this whole thing is hydrogen storage onboard. Everybody is talking about hydrogen fuel cells and batteries on them, but they've all been done before. What has not been done before is taking cryogenic hydrogen onboard, which gives you the right range. That's where we are working with some labs in cracking that problem. We already have a couple of solutions in hand when you're trying to make it more efficient. And that's where we're using our grant money. The primary advantage of hydrogen is the length of flying time.

"Some of these companies that are already flying with batteries take a long time to charge and they have to land quite frequently to charge. The ranges are also limited and that's going to be a logistical nightmare and a high cost of operations. I don't want my vehicle to cost $5 million, I want them to be affordable and hydrogen makes it affordable because you can have a long range. The cost per mile will be low with hydrogen compared to batteries. We just want to make it operationally efficient so it can reach a larger population. We want to democratize the larger population and are not going to make it exclusive, serving a small community. So that's where we are. The hydrogen engine we are developing has its own battery management system. We are making our own hydrogen fuel cell engine and hydrogen is being developed in partnership with Bosch Aviation, and with a few other fuel cell companies. That is something we're developing exclusively for LuftCar. But it can also be custom built for other eVTOL companies. We will sell them to other eVTOL companies as well, so that's another revenue stream. We will hit the market with revenue for that for sure.'

Whether a flying car or a car that is being flown, these and other electric aerial vehicles still need facilities and support systems when they are on the ground, which we discuss in the next chapter.

6 GROUND OPERATIONS

The Rise of Vertiports

While much of the discussion in this book is about the flying activities of the new electric air vehicles, there's a lot that has to go into the infrastructure for the handling of the vehicles while on the ground. While most EAVs don't require runways, since they take off and land vertically, they do require specific locations to land, change or charge batteries, pick up and drop off passengers, and be serviced. These functions would be performed at Vertiports. As the name suggests, these are facilities for vertically operating vehicles, much like heliports are used for helicopters.

Vertiport developers face a host of major and minor issues, such as local restrictions, building permits, considerations of things that are nearby, meeting FAA guidelines, the potential amount of air traffic, nearby restricted airspace, and countless other issues. In some cases, heliports are being converted to accommodate EAVs. In other cases, they are being outfitted at both larger and smaller airports. And yet in other cases they are being developed as standalone facilities outside of traditional airports. Many of the facilities are designed for the rapid turnaround of EAV takeoffs and landings to accommodate coming air shuttle services, such as between airports or between airports and designated city or town locations. Because the electric passenger-carrying vehicles will be flying in and out of vertiports and in many cases over populated areas, expect the facilities to be highly regulated with a great emphasis on sound safety.

The good news is that vertiport infrastructure developers have experience in airport infrastructure development or operations with safety being part of their corporate DNA. Everyone I interviewed in the course of research for this book brought up the issue of safety, noting its significance throughout the entire industry. As you can see from the following description of vertiport developers around the word, there is a rich heritage of dealing with vehicles going from ground to air and back.

Skyports Infrastructure

Skyports Infrastructure in London has an autonomous drone service used for cargo delivery services and survey and surveillance operations. For example, Skyports Drone Services launched a program in Yeosu City, Korea, to deliver pharmaceuticals and food by drone. Skyports and drone partner RigiTech launched flight trials by a drone service consortium including Skyports Drone Services Korea, Marine Drone Tech, and RGB Lab.

Skyports also designs, builds, operates, and sometime owns, vertiports for eVTOL vehicles globally. The company's design for a vertiport in Dubai was approved by Sheikh Mohammed bin Rashid Al Maktoum of Dubai with a planned

2026 launch. In 2022, Skyports Infrastructure and private airport operator Corporación América Airports S.A signed a memorandum of understanding for the plan and development of vertiport infrastructure by sharing their best practices for vertiport design concepts in Latin America. That partnership included developing vertiport networks planning for the deployment of permanent vertiports.

Urban V

Urban V in Rome has established Aeroporti di Roma, SAVE Group, Aeroporto di Bologna and Aeroports de la Côte d'Azur to expand vertiport operations starting with Rome. In true paired interest form, Urban V partnered with global aviation consultancy Mott MacDonald to develop advanced air mobility infrastructure. Urban V plans to connect Fiumicino International Airport with central Rome for eVTOL travel by the end of 2024. Urban V also has a partnership with eVTOL jet maker Lilium to develop vertiport infrastructure for Lilium aircraft and passengers with an initial focus on Italy and the French Riviera. In 2022, Urban V, Aeroporti di Roma, Volocopter, and Atlantia conducted the first crewed, eVTOL's public test flight in Italy along with presenting the country's first advanced air mobility testing vertiport.

Ferrovial Vertiports

Ferrovial Vertiports in Austin, Texas, is a division of Ferrovial Airports. Parent company Ferrovial is a 70-year-old infrastructure company that builds airports, highways, hospitals, schools, and various other buildings. In 2021, Ferrovial Vertiports and eVTOL jet maker Lilium signed a framework agreement to develop a network of 10 vertiports in major cities in Florida. In 2020, Ferroviial launched its Foresight digital platform for the sharing of innovation trends, knowledge, and new technologies with clients, startups, and venture capital firms. The platform included features for scenario planning and detailing industry trends. Ferrovial's airport division also partnered with U.K. real estate developer Milligan to develop locations for vertiports in the U.K.

Groupe ADP

Groupe ADP in Paris is an international airport operator that manages more than 25 airports directly or with TAV Airports in Turkey or GMR Airports in India. In 2020, Groupe ADP partnered with urban transport operator RATP Group to develop an urban air mobility industry branch with a focus on eVTOL crafts. Volocopter, the Germany based electric air taxi company founded in 2011, was chosen for testing in the Paris region with Groupe ADP, which started in 2021. At the International Paris Air Show in mid-2023, Groupe ADP said the Paris region would be the first European city to offer electric vertical take-off and landing aircraft services in time for the 2024 Olympic and Paralympic Games. By late 2023, numerous local authorities opposed the launch of the new air services due to environmental concerns.

Volatus Infrastructure

Volatus Infrastructure and Energy Solutions is a key cog in the evolving electric aerial ecosystem. The company can be commissioned to develop vertiports but it also partners with other companies to help develop their particular infrastructure.

For example, Volatus Infrastructure signed a memorandum of understanding with Australian flying car maker Ace VTOL to be its infrastructure partner. The deal would help enable Ace VTOL to expand its business in the growing eVTOL market in Perth, Australia. Volatus Infrastructure also is working with Japanese eVTOL maker SkyDrive to work on an advanced air mobility infrastructure in South Carolina, where SkyDrive is establishing its home base. It also is a partner for Aeroauto, which offers vertiport construction as an extra service outside of its EAV showroom business.

Volatus Infrastructure has ongoing vertiport projects at Bellefonte Airport in Pennsylvania, Greenport International Airport in Austin, Texas, and Whittman Field, in Oshkosh, Wisconsin. The company segments its ground operations systems into four approaches.

- Vertiports – Includes the landing infrastructure, charging station, and a terminal building following FAA rules.
- Vertihubs – Larger vertiports that also can provide maintenance, repair and overhaul services.
- Vertistop – Provides the ability to land and recharge vehicles, more of a bare bones approach. The FAA has guidelines for these minimally designed vertiports.
- Landing Pad – These are aimed at single–person, light aircraft, facilities that do not require federally compliant infrastructure.

Bob Johnson is Chief Commercial Officer of Volatus Infrastructure and oversees business development, making sure the company has the right strategy and relationships to help scale the business.

"We formulated about three years ago and became legal about 18 months ago," Johnson tells me in an extensive interview.

The company is based in Neenah, Wisconsin, the location of the world-famous Oshkosh air show held at the Wittman Regional Airport. More than 650,000 people attend the annual show with more than 10,000 aircraft flying into Wisconsin for it. The event is run by EEA (experimental aircraft association) Ventures, a perfectly fitting location for a vertiport startup.

Getting Started

When Volatus Infrastructure started a few years ago, EAVs were in development but not yet on the public radar and the FAA had not yet started issuing any significant numbers of flight certification approvals for such vehicles. The company decided to get a step ahead of the curve and start planning for managing electric aerial vehicles when they were on the ground, a significantly more complicated mission than it may seem.

"We looked at the initial let's do something that no one else is doing," says Johnson. "We look at it from an approach that we want to be pragmatic. We want to be simple. We want to be affordable, and we want to be efficient. If you look at a lot of the other guys that are building vertiports, they're going into these large airports throughout the world. And those vertiports must be something in the neighborhood of $10 million to $20 million. We're not selling on price; we're selling on value and we're selling to scale. I'm not trying to compare prices here. But if I can put in a vertiport that does the job for $750,000 to a $1 million, versus $10 million to $20 million, what can you do with that other $9 million for infrastructure?

"There's an engineering Brief 105 (see Appendix) that the FAA has put out, and that's going to be more or less the blueprint for the concept of operations. Of course, there will not be an official ruling on that from the FAA until probably 2025. Certifications are going on with Joby Aviation, Archer Aviation, Wisk Aero, and other equipment manufacturers throughout the world. And it probably won't be until 2025 that we get these aircraft in the air. A lot of people wonder where the money is coming from, but we've got to have a certified aircraft before people are really going to take the infrastructure seriously. We understand that, but it's going to start at existing airports. The reason for that is they're already approved. They already have an electric grid. The general public feels safe knowing that that works."

The Blueprint

Once deciding to enter the business of developing vertiports, Volatus Infrastructure had to create a blueprint for the road to create vertiports, an entirely new business infrastructure to serve an entirely new business category.

"First, you need to do a feasibility study. And you need to find out where the people are. And second, you need to get a demand model. If we were to do this, how many people are in this area? What are their backgrounds? How much do they spend and what kind of modes of transportation do they do? You look at nature. Is this close to water? Is it close to a swamp? Do they have birds? We don't want bird strikes. Can we get electricity out here? Can we build a substation? How much is it going to cost to build that substation? What about the environment? What about permitting? What about zoning? Can we do that? We have a lot of effort. We have to look at the buildings. What is flying over us? What obstructions are there? Is there an airport over here? Is there a flight path over here? How many people are we going to be flying over? And you go wow, there are a lot of a lot of these. They're going to make our industry. We can't live without them. But commercial real estate needs to understand that we cannot build a vertiport on every one of their buildings. They think that even in California, the vertiport infrastructure on existing buildings is going to be a challenge because of the earthquakes."

The economic realities of operating a vertiport also are critical. This is a classic chicken-and-egg situation, with the electric flying vehicles not yet flying commercially but there is no on-the-ground infrastructure yet created for when they are ready for business. An additional challenge is the scope of a business. As in any technological startup situation, starting small and then growing is the story of the day. That means a vertiport business has to live in the 'build it and they will come' world for a while.

"You've got to have throughput at these vertiports. In a central business district of Los Angeles, or Atlanta, or New York, you're looking at maybe four- to eight-bay terminals," says Johnson. "Some are putting double-decker levels on the terminal. As we say in the train business, if the train isn't running it isn't making business, so you've got to have throughput. We're looking at probably two to three flights per hour. And you've got to keep in mind that you've got to charge those. The question is, are we going to be swapping batteries, are we going to be bringing the charger out to the eVTOL? How many volts is that going to be? How long does it take to charge that? How can we get that approved with the communities involved?"

Community Backing

Local and regional communities have not yet weighed in on how receptive or opposed they will be to local takeoff and landing facilities being located within any given area. When telecommunications companies started rolling out towers for serving the growing transmission capabilities needed for rapidly evolving cellphone technologies, they quickly found that not everyone wanted a tower built anywhere near them. In the case of electric aerial vehicles flying around at relatively low altitudes, taking off and landing at regular frequencies, and rideshare services coming and going to these landing facilities to pick up and drop off passengers, local community relations promises to be a significant coming market need. Developing a vertiport faces numerous challenges, but they are a critical component for the electric air future.

"We see the end of the road and we're not giving up. We're in this full bore right now. But you've got to be safe. And you've got to make sure that you know there are a lot of batteries out there that are catching fire these days. That is a really, really big challenge. We're looking at chargers that have intermodal agnostic type connections. We're talking to people about hydrogen. And now that's down the road. But I think the way we're making hydrogen now is heating up natural gas. We've got to look how to split the molecule now and be able to get that power."

Since the electric aerial revolution becoming widely public is so new, expectations will have to be managed. When people hear of a new vertiport coming where relatively quiet electric vehicles can come in and quickly whisk them to desired destinations, they may self-create unrealistic expectations. The innovators creating the flying and ground infrastructure are realistic in the required efforts all around, since many have been working on it behind the scenes for a number of years. While the electric aerial revolution has been in the works for years, it is very new to the general public. Volatus Infrastructure is keenly aware of the market complexities. Despite the many challenges, the company has a strategic and tactical approach to getting its business off the ground.

"Basically, we get a location, and we work with civil engineers," says Johnson. "From our standpoint, we are modular. We will be able to ship a vertiport in a container or couple of containers that can be assembled without heavy machinery. That's what we're looking at being pragmatic because we've got to walk or crawl before run. When you see these first vertiports, some people are going to be taken aback because they're expecting to have a Starbucks there, a TGI Fridays, their massage, whatever. And that isn't going to happen, because the point of a vertiport, in our in our perspective, is for convenience. We want to get the cats in there, get the cats out. We're herding cats and that's hard to do. TSA is one of the biggest holdups, as we know."

Planning for Scale

At its start, Volatus took a strategic look at the overall aerial market and foresaw the coming large deployment of electric vertical takeoff and landing vehicles. The company identified what it saw as coming needs and se its focus there. It had a view of a grander picture than just a few aerial vehicles going from here to there in limited fashion.

"We were thinking at the time there was a need for all these beautiful electric vertical takeoff and landing airplanes," says Johnson. "And this electric revolution, where they use electric power distribution, which is much more efficient than the gasoline engines you see today. We saw that there was a need to make sure that these airplanes and eVTOLs had a place to land and take off. Not many people really thought about that. You always think about the Ferraris and the beautiful cars and the sexy cars and all these renditions on the internet of how beautiful they look. But you've got to do things that are dull, dirty, and dangerous. And maybe that's what we're doing here. We figured that we've got to have a place for them to get the juice, get the electricity to take off, and make sure that that's legal. Make sure that people know what that is, make sure that people accept it, make sure most of all that you're safe and you're compliant. And at the end of the day, let's make sure you have the throughput, and you can scale. And you can grow a business and grow a sector and save lives and change the world."

The approach of Volatus Infrastructure was rather straightforward: If electric vertical takeoff and landing vehicles were going to exist, which led to the obvious conclusion that they were, then there would be a need for landing, takeoff, and electric charging facilities. That rationale may sound simple, but it was a relatively small number of people and companies that saw the wave of electric aerial vehicles coming and were thinking along those lines.

"Landing places were needed," says Johnson. "Right now, we're going after the existing infrastructure. But if you've noticed in the past few months, or a couple of years, there's a lot of interest in personal flight vehicles. You have written about this. And if you look at where the FAA is compared to the other regulatory countries, with the FAA, we have the safest national airspace in the world. There's a reason for that. But we also know that this is a brand-new sector. We don't have a lot of data in order to make things safe. It's going to take a little bit longer to make sure that we continue to have the safest national airspace system in the world. That's where the idea came from. We decided this would be something that a lot of people aren't doing. You want to do things that others don't want to do, and you want to differentiate yourself. That's where we came up with the idea."

Starting at Airports

While it has four distinct systems for creating takeoff and landing facilities for electric aerial vehicles, the company is taking a pragmatic approach to align with the market. The FAA has suggested that the best place to operate the electric air vehicles in the beginning is from established airports, which already have the infrastructure for traditional planes, including established communication and emergency procedures. Volatus is fully aware of global airport infrastructures, the competitive market landscape for vertiport constructions, and the FAA desires to start commercial enterprises at airports, even though Volatus has the ability to approach projects in a smaller, modular fashion.

"We're doing both," says Johnson. "We are in talks with the FAA and with Oshkosh Airport, which is Wittman Regional Airport. There, there are processes for airport layout plans that you've got to go through with the U.S. Department of Transportation and the state Department of Transportation to get approval. There's a lot of a lot of I dotting and T crossing in order to do that. We're also working with private landowners as well. We have signed a deal with Belafonte Airport in Pennsylvania. We've also signed a letter of intent with Greenport International Airport, which is going to be the largest private airport in the world. It is built just outside of Austin. And we're very excited that we were signed on to be the vertiport provider for that. It takes up the entire Bastrop County.

"It's going to be very interesting, because they're going to be the first green, carbon net zero airport in the world. It's also going to be an international port of entry and it is going to be private. It's going to complement Austin International Airport as well. They're going to have in the neighborhood of three million square feet of hangar space, some ridiculous amount of hangar space, because there are going to be a lot of vehicles there. There's no doubt it's going to take years to build, but you have to start somewhere."

While Volatus Infrastructure is focused primarily on larger commercial operations, it recognizes that 2024 is the year the powered ultralight category of personal aerial vehicles are hitting the market. These single-person vehicles do not require any significant takeoff and landing infrastructure and can be flown from personal property.

"The ultralights are coming out in a big way," says Johnson. "What we're going to see naturally is in the Paris Olympics that will be happening in France. And you'll have Volocopter over there. There are rumors that President Macron will be one of the first passengers on the Volocopter and they're going to have five routes. That's going to be the wake-up call to the world that eVTOL is here. You'll see that 2024 will probably be the year next to shut up or put up. You're going to see a lot of companies that have been burning through cash to build these things and you're going to see them probably run out of cash. A lot of companies will consolidate and consolidate their IP, and you're going to see a few real players, within the next probably 10 to 15 years. You're going to have three to five major OEM players. And there's some that have not even built their prototype yet. Some will take a wait-and-see attitude and let other people make mistakes. Let's understand from them, and let's understand how to complement the transportation sector, they are a part of that."

Flying with Pilots

The market comprises three major segments, as discussed earlier in the book. The one-person ultralights, the certified personal crafts, and the commercial transportation categories. While the ultralights do not require a pilot license to fly, the FAA certified vehicles require a pilot, at least for most at the moment. Although untested in market perception, some electric aerial vehicles are being designed to run pilotless. For example, a Chinese eVTOL vehicle maker received permission to fly passengers in its unmanned aerial vehicle. EHang Holdings received the type certificate (TC) officially issued by the Civil Aviation Administration of China (CAAC), and in late 2023 received a standard airworthiness certificate (AC) from the CACC, making it the world's first unmanned eVTOL two-passenger vehicle cleared for commercial use. In May 2024, EHang flew its first passenger in a demonstration ride in Abu Dhabi. The vehicle had flown several unmanned flights during the DriftX Mobility Expo in Abu Dhabi in April 2024. Johnson expressed his views on pilotless EAVs and how the market is likely to evolve.

"Today they're piloted, of course, and hopefully, within the next few years, it will be autonomous," says Johnson. "Initially, you will see cargo and EMS take hold before you see passenger travel. And the reason I say that is for safety reasons. If you look, I think we'll be flying boxes over rattlesnakes, over deserts, over water at nighttime. If there's a mishap, we can learn from that. There's a lot going on right now with that industry. But that's where you're going to see it. You're going to see a lot in the other corners of our world. Specifically, in Southeast Asia, Australia, a target area for personal vehicles, we are very interested in in building vertiports. For those guys, we understand that it will be private landowners. And sometimes being in the position that we are in could be what I would call a conflict of interest. And we're in the eVTOL world with the big guys, with Joby and Wisk and the transportation sector. What we want to do is democratize the era, and we want to make it affordable for everyone. And not just for the affluent and the rich.

"If you go back to when we started flying jets, if you look at all the pictures on the internet, you'll notice that wow, these guys were dressed so nicely everyone had suits on and they were serving all this fine food. And now there are people wearing tank tops and tennis shoes and all that. That's because the affluent could afford those prices back then and that's why you didn't see that. We want to make sure it's this sector that we go forward with that and make sure that we can make it affordable for everyone and we complement the modes of transportation."

Ticket Packaging

Major airlines are highly involved in the electric aerial revolution, both to protect their core business and to expand it. There was a time some years back when there was linkage between shorter-and-longer term travel. For example, if I was flying full-fare first class from JKK Airport in New York to London, I would receive a complimentary helicopter service from the Manhattan heliport directly to JFK. Johnson sees similar airline packaging down the road.

"All of us as a sector are working on an app where if I want to leave my house, we would use the app," says Johnson. "I'm in Atlanta, and I fly out of Hartsfield-Jackson International Airport, the world's busiest airport. If I want to go to San

Francisco, I'll be able to pull one app up on my phone. It will get me an Uber from my house to the vertiport, the vertiport vehicle would take me from home in Buckhead or Sandy Springs, Georgia, down to the airport. Then I will fly to California, and it will take me to my location. That is all in one app, and you pay one price. That's the ultimate goal that we're trying to do. With all that in mind, when you get to that vertiport, you would already be approved, because we will know basically how many hairs you have on your head. We also will know approximately what your baggage weight will be for the airplane for the entire trip.

"United Airlines is a member with Archer Aviation and Joby Aviation has partnered with Delta Air Lines in Atlanta. Joby Aviation is doing what they call a home-to-seat. Basically, it is for their preferred customers. We're going through some testing right now, their premier service, of course. But that is what a lot of the airlines are doing. You also need to think about why the airlines are doing this. United has cut a check to Archer for $10 million. They said they sold 100 of Archer's eVTOL aircraft, which are being built in Covington, Georgia, outside of Atlanta. We now have a pilot shortage, and as you are well aware of, we have maintenance, repair, and operations. A lot of the helicopter pilots are now going to the airlines. And we're asking who's going to fly these the eVTOLs and how do we train these guys? How does the FAA get into training? And how did they get their pilot's license? Is it 1500 hours? I was at a meeting here in town with the FAA and they came in and said they just had a meeting with about 30 fourth graders and we don't know if these fourth graders want to get into aviation. But we've got to start them early. And when you are able to say you can use your flexible fingers that young people can, it's the same thing as finding a stick. If we can get that generation excited about that, make sure that the pay is equivalent to what job they do. Being a pilot could be a part time job. One good thing about it is you will probably be at home every night in your bed, so you're not going to be on these weekly furloughs and 10-day furloughs that a lot of the pilots are today. So that's an advantage. There's a lot to be worked out for that. But we think simulation is where it's going to go."

Loading Facilities

The actual vertiport will have some facilities beyond a location for the vehicles to take off and land. Between flights, batteries need to be charged or changed. Passengers need to get on and off the vehicles. There must be some kind of security, both for the vehicles and the passengers. There needs to be waiting areas and facilities for passenger drop off and pickup by land vehicles, if the vertiport is not at the final destination.

"At vertiports, you're going to have a charging station," says Johnson. "You're going to have a lot of technology that you don't see. You'll have technology that says, can we fly here? You will probably have a coffee shop. You're going to have refreshments, but very limited at first. You'll have parking. We want you to be on the air side. On the air side means that you've already passed the gate on your way needing to get on the plane. You're going to have people that are remote pilots that are on the ground. We don't know if there'll be in the actual airport, they are at the vertiport, they probably will in the beginning. When Wisk flew their generation five in Oshkosh a couple of months ago, it was autonomous. It was being controlled out of California. It doesn't really matter where you are. You just must have connectivity.

The vertiport infrastructure market is so new, the best models and end game are yet to be determined. Some will come from traditional airport operators, as discussed earlier in this chapter, some will come from the actual makers or operators or electric aerial vehicles and some will come from vertiport specialists, such as Volatus Infrastructure.

"Right now, there's a mixed bag of operators, builders, service providers, vertiport providers," says Johnson. "You've got some people that want to do the whole ball of wax. They want to build it, they want to operate it, they want to fly it. They want to build their own infrastructure. But some people say they want to build it. And we want to sell our airplanes and let these fixed-base operator type providers come in and do the service. I think the reason that people like Joby and Wisk and Archer are saying, at first, we want to fly our own product, in case something happens. Some equipment manufacturers think it will be better for us to build this OEM."

A Global Approach

Like numerous other startups in the electric aerial vehicle market, Volatus Infrastructure views the world as its market. While activity in Asia and the Middle East is moving aggressively ahead, the market in the United States can sometimes be a bit more complex due to numerous stakeholder parties.

"In a country like the United States, we've got 50 decision makers on how we're going to do the airspace and how we're going to verify everything," says Johnson. "You've got other countries, and cities like Singapore, and Dubai, and you've got central governments. From a rulemaking perspective, that's going to be a lot easier to get these things up in the air. That's where I think you're going to see a lot of the first flying, that's going to happen. So kudos to those guys."

Many companies throughout the business of electric aerial vehicles need to precisely target their sweet spots. The coming market is so big, with totally different categories of flying vehicles and widely different regional markets globally, it can be challenging to pick a spot and stick with it.

"You've got to focus on your core, you got to focus on your mission," says Johnson. "I see a lot of people today in this sector, both infrastructure and even flying, at what they call boiling the ocean. You've got to limit the scope of what you want to do. And you can't be everything to everybody. But it's hard not to say if you've got a big oil company coming in and saying we would like to talk with you about what you're doing, we might want to talk to you about hydrogen, it's hard to turn those people down, especially when you're raising funds. They can eat you up and spit you out. That's the toughest challenge that a lot of us have right now.

The electric aerial business is not so much about travel as it is about overall logistics. It's about moving people and things totally differently than in the past and providing the movement of people and things to and from places differently from the past.

"You hit the nail on the head earlier in the conversation, you talked about logistics,: says Johnson "If you look today, there's a company out of England, Dowty Propellers, a GE Aviation company that makes propellers. Over the last 25 years, they've made maybe something like 69,000 propellers. If you look at the forecast of the propellers that are going to be made, it is in the millions. And it's like I think a press is what makes the propellers and a press to make that cost $15 million to $20 million. And there are not a lot of presses out there today. I think you're going to see a mad scramble, when these things start going. And it's like, we didn't think of that. So that's what we're trying to do. I've been an early adopter in disruptive technologies all my career and I've been doing it for 43 years."

Working Together

Volatus Infrastructure is but one example of a startup company looking to find and cement its place in the gigantic evolving world of electrical aerial aviation. It's way too early to measure the success or failure of any of these companies since they all are looking to provide their own value component to the overall matrix.

"What would make us succeed is that everyone succeeds," says Johnson. "We're not competing with anyone in this industry, and I will be the first to say that. We want to complement our competitors if you can call them competitors."

The reality is the market is highly complex with many moving parts, both from a technological and a regulatory and safety standpoint. Companies are navigating all in their own part of the overall ecosystem.

"It's very simple that a lot of the business plans say we want to be a one-stop shop," says Johnson. "It's like I want one person that I can go to. And I can say I want to buy this vehicle, I want to know what the warranty is, I want to know how I get the chargers here, and I want the airspace done. There are a lot of moving parts in this."

Volatus Infrastructure is working with various departments of transportation. The company is working heavily in creating partnerships and becoming established with the right people from the infrastructure side and from the equipment manufacturing side. The company is realistic about what it and other companies in this arena face.

"Fear is my biggest number one concern because of the fear of even building cell towers years ago. You're familiar with not in my backyard (NIMBY). The toughest part right now is making sure that we get these demos out."

The key for all these startups is getting it and getting it right. That is where Volatus Infrastructure has all its focus at the moment. Vertiports also will be needed for the new electric air taxi businesses coming, which we discuss in the next chapter.

7 FLYING TAXIS

The New People Movers

In all the business aspects relating to flying vehicles, the largest scope relates to flying taxis, which fall into the category of CAV (commercial aerial vehicle). These are planned around the world and are intended to shuttle passengers between locations that may be too short for a traditional airline flight but a bit long for a practical drive. Some of these are vehicles designed to carry one person along with the pilot and others are intended to carry multiple people at a time. Many of these vehicles have been in development for years, as is the case with other categories of flying vehicles. Many also have received FAA certifications for flight testing and have been very active in trials around the world, a pace continuing through 2025 on the road to commercial passenger travel in 2026, a timeframe guideline of the FAA.

Flying taxis are a key part of the low-altitude economy, discussed earlier in the book. As an integral part and due to the scope of operations, there is massive investment involved. The cost and resources needed to develop large-scale flying taxi vehicles and services are immense. Flying capabilities must be determined and validated, generally conducted initially without pilots on board, since the vehicles can be flown remotely, similar to drones but on a larger scale. Flying prototypes need to be created to introduce the concept to various markets around the world, since this is a totally new industry. Takeoff and landing facilities need to be arranged and, in many cases, constructed. For airport operations, systems have to be integrated with traditional flight control systems and in the context of flights by commercial aircraft. Systems for passenger pickups and drop-offs need to be designed and battery charging facilities established. All of this takes considerable time and resource, one of the reasons widespread services are not expected before 2026.

The business of flying taxis could be developing, manufacturing, and selling the flying vehicles. In other cases, the business could be running the flying taxi service. There are several major flying taxi companies, most extending their reach around the world. Various branches of the military also are exploring and testing these since the vehicles can transport people as well as cargo with less stringent airport requirements, such as runways and fueling depots. The vehicles also are quieter than helicopters and considered safer. The U.S. Air Force has agreed to purchase six flying vehicles from Archer Aviation in a $142 million deal.

Efficient Travel

Flying taxis are about increasing the efficiency of travel. For example, a flying taxi could transport passengers from Newark Airport in New Jersey to New York City in 10 minutes or from Heathrow Airport in London to downtown London in 15 minutes, avoiding typical traffic bottlenecks. Flying taxis are expected to replace 90-minute drives with 20-minute flights, The Midnight flying vehicles from Archer Aviation can take off vertically, fly at 150 miles per hour, and carry four passengers along with luggage. Flying taxis will be good as shuttle services, since vehicles like the Archer Midnight can do back-to-back, 20-mile flights with 12 minutes of battery charging time between flights. Initial passenger costs per trip are expected to be in the range of a top-tier ride share service.

Flying taxis are a totally global phenomenon and a very large-scale proposition. What is coming is a total upheaval in how people move between locations, leading to a transformation of where people go and how they get there. The coming market involves thousands of flying vehicles operating at low altitudes, shuttling people to and from. The long-term future will impact land travel, including road-building requirements and automotive needs, as moving transitions to the air. Flying vehicles will evolve to be self-flying, leading to changes in how people around the world move. The electric aerial vehicle from EHang Holdings in China already has a pilotless flying vehicle and plans pilot-free air travel. While longer-distance travel will continue in the high-altitude economy of traditional aircraft, many long car drives will transition to short flights.

Large Scale Approach

Some air taxi approaches are very large scale with very large ambitions, and with good reason. Establishing a worldwide flying taxi business from ground zero takes a lot of planning, research and development, testing, and funding. A good example of such scope and scale is Archer Aviation. The Santa Clara, California, startup is backed by United Airlines and Stellantis. United paid Archer $10 million as pre-delivery payments for 100 of 200 EAVs covered under a purchase agreement. In 2021, Archer entered into a purchase agreement with United worth up to $1.5 billion in eVTOL sales. Archer already has plans to start service in the United States. In 2022, the first commercial electric air taxi route from Newark Liberty International Airport in New Jersey to Downtown Manhattan Heliport was announced, promising to convert a one-hour drive to a 10-minute flight. United is clearly all-in on eVTOLs, with Scott Kirby, United Airlines CEO, noting: "Archer's eVTOL design, manufacturing model, and engineering expertise has the clear potential to change how people commute within major metropolitan cities all over the world."

Archer has been working on its flying vehicles operations since 2018. By 2021, Archer had strategic relationships with United Airlines and the United States Air Force and had completed its first test flight, along with the conceptual design of its Midnight flying vehicle. Archer has other arrangements as well, including a memorandum of understanding to sell Air Chateau International, a leading private aviation operator in the United Arab Emirates,100 Midnight CAVs valued at up to $500 million. Air Chateau plans to operate the Midnight eVTOLs in the region. Air Chateau' plans to offer last mile services to ultra-high net worth individuals, with connectivity between airports, cities, strategic points, and key attractions. Archer Aviation also has an agreement with the Abu Dhabi Investment Office (ADIO) to work together to make Abu Dhabi its first international launch partner, with plans to start air taxi operations in Abu Dhabi in 2026. ADIO plans to provide Archer with incentives to establish its first international headquarters and manufacturing facilities in the Smart and Autonomous Vehicle Industry Cluster in Abu Dhabi.

Expanding Geographically

The air taxi plans of Archer Aviation continue to expand beyond the United States and the United Arab Emirates. In India, InterGlobe Enterprises, a large travel and hospitality conglomerate, signed a memorandum of understanding with Archer Aviation to partner to launch and operate an all-electric air taxi service across India in 2026. That arrangement includes the financing and building of vertiport infrastructure with plans to finance the purchase of up to 200 of Archer's Midnights for the India operations. Other uses of the EAV may include logistics and medical and emergency services.

To manufacture its flying vehicles, Archer closed agreements with Synovus Bank and Evans General Contractors, LLC, for the financing and development of a high-volume manufacturing facility in Covington, Georgia. The facility is expected to eventually be able to produce more than 2,000 EAVs a year, making it the world's largest eVTOL manufacturing facility by volume.

For pilotless travel sometime in the future, Archer agreed to make Boeing subsidiary Wisk its exclusive provider of autonomy technology for future variants of Archer's EAV. As part of that collaboration, Boeing agreed to make an investment in Archer to support the integration of Wisk's autonomous technology in future pilotless CAV, which we discuss in the next chapter.

Years in Development

While the idea of flying taxis may not be common knowledge, many of the development efforts have been going on for years. Joby Aviation, in Santa Cruz, California, has been working to develop its flying vehicle for more than 15 years. Even before Joby Aviation, CEO JoeBen Bevirt founded Joby Inc., which created the well-known Gorrillapod flexible camera tripod (one of which I still own) and Joby was eventually sold to DayMen Group. Joby Aviation was started by Bevirt in 2009 with $100 million backing from companies including the venture capital arms of Toyota, JetBlue, and Intel. Toyota is Joby's largest external shareholder, having invested approximately $400 million in Joby since 2020. The two companies signed a long-term agreement for Toyota to supply key powertrain and actuation components to produce Joby's CAVs. By late 2023, Joby had $1.1 billion cash on hand.

Major airlines are highly active in the eVTOL marketplace, as discussed earlier in the book, and Delta Air Lines has been active with Joby. Delta is working with Joby on a multi-year, multi-market commercial and operational partnership to deliver home-to-airport transportation service to Delta customers, beginning in New York and Los Angeles. The two companies are working to integrate a Joby-operated service into Delta's customer-facing channels, providing customers who travel with Delta through New York and Los Angeles the ability to reserve a seat for short-range journeys by EAV to and from city airports when booking Delta travel. Delta made an upfront equity investment of $60 million in Joby in 2022, with the option to expand the total investment up to $200 million as milestones are reached. The premium Delta service would run alongside Joby's standard airport service in priority markets. The partnership is mutually exclusive across the U.S. and U.K. for five years following commercial launch, with the potential to extend the period. The initial launch is expected to be eVTOL service from New York's JFK International Airport to the New York City heliport in a seven-minute flight. In late 2023, Joby performed an exhibition flight in New York City, the first electric air taxi flight in the city and the first time Joby flew in an urban setting.

The Joby eVTOL has six electric motors and can carry a pilot and four passengers. It can reach speeds of 200 miles per hour with a range of 150 miles. Joby plans to operate an all-electric, vertical take-off and landing vehicle with fast, quiet, sustainable, city-to-airport service by air. Joby's CAV full-scale prototype already has logged 30,000 miles flown.

Production Mode

Like Archer Aviation, Joby Aviation is geared for flying vehicle production. Joby selected Dayton, Ohio, the birthplace of aviation and home to Wright-Patterson Air Force Base. for a large-scale aircraft production facility to deliver up to 500 EAVs a year. The site is large enough to create up to two million square feet of manufacturing space and is expected to be up and running before 2026. In 2023, Joby made its first delivery to the U.S. Air Force as part of a $131 million contract with the Department of Defense at Edwards Air Force Base in California

As other flying taxi companies, Joby is taking a global approach. The company formally applied for its EAV design to be certified for use in Japan, with Japanese and U.S. regulatory authorities agreeing to deliver a streamlined approval process for U.S. applicants to validate their eVTOL designs in Japan. In 2023, Joby partnered with ANA Holdings and Nomura Real Estate Development Co., one of Japan's largest real estate developers, to develop take-off and landing vertiports to support the commercialization of electric air taxi service across the country. Joby also signed an agreement with SK Telecom, who invested in Joby in 2023, to jointly undertake flight testing in South Korea in 2024 as part of the Korean government's desire to support the development of the Urban Air Mobility sector. In early 2024, Joby agreed with the Dubai Road and Transport Authority to launch an air taxi service in Dubai with initial operation starting in 2025. The agreement gave Joby exclusive rights to operate air taxis in Dubai for six years starting in 2026.

Flying taxis are meant to fit in with regular airport traffic and work within the current aviation network. To model how that could work, Joby conducted a series of air traffic simulations with NASA's Ames Research Center to evaluate how air taxi operations can be integrated into airspace including at busy airports, using existing air traffic control (ATC) tools and procedures.

The simulations, jointly developed by Joby and NASA airspace engineers tested scenarios with dozens of eVTOL aircraft per hour flying into and out of the busy airspace in the Dallas-Fort Worth (DFW) region using the current airspace system.

During the simulation, participating teams of controllers virtually tested the ability to integrate up to 120 EAV arrivals or departures per hour from DFW's Central Terminal Area, alongside the airport's existing traffic. Up to 45 simulated eVTOL aircraft were safely simultaneously aloft in DFW's Class B airspace during the activity.

Partnerships

Flying taxi operations around the world often involve partnerships, primarily due to the complexity of operations, the large number of stakeholders, and the required resources. One good example of this was Volocopter, a 10-year-old German company that created partnerships around the globe for its eVTOL flying taxis. Volocopter has a trio of flying vehicles: VoloCity Air Taxi, the VoloRegion long-haul passenger vehicle, and the VoloDrone heavy-lift cargo drone. The two-seater VoloCity has 18 propellers, nine batteries, and a 17-mile range. While the company is headquartered in Germany, it has been conducting flight demonstrations of its Air Taxi in cities around the world. However, flight demonstrations are just a light sampling of the depth of Volocopter's partnerships.

In Italy, Aeroporti di Roma, Atlantia, Urban V, and Volocopter completed the first crewed eVTOL test flights in Italian airspace. The tests were part of a mobility ecosystem setup at Fiumicino's Leonardo da Vinci International Airport, a key milestone toward the planned rollout of advanced air mobility services in Rome in 2024, with the partners having initiated operations of the first fully functional vertiport in Italy. In Canada, CAE Montreal, a leading aviation training organization, and Volocopter signed a strategic partnership to develop, certify, and deploy a pilot training program for eVTOL operations. In France, Groupe ADP, Skyports, and Volocopter commissioned Europe's first fully integrated vertiport terminal for the urban air mobility industry at Pontoise-Cormeilles airfield. The facility allows advanced testing of critical technology and passenger processes.

In Germany, Volocopter and SITA, an air transport industry IT provider, entered into a partnership, with SITA selected as Volocopter's preferred digital and IT systems partner for vertiports. In Saudi Arabia, NEOM, the smart regional development in northwest Saudi Arabia, and Volocopter completed a series of air taxi test flights, the first time an eVTOL vehicle received a special flight authorization and performed test flights in the Kingdom of Saudi Arabia. In China, since announcing Geely Technology Group strategic investment in Volocopter in 2019, the companies have been using their combined expertise in mobility to identify where electric air taxi services could be best introduced in China.

In Japan, the global trading company Sumitomo Corporation invested in Volocopter and became a strategic partner for entry into service in the Japanese market starting in 2025. Despite all the forward progress, the continuing need for funding drove Volocopter into insolvency. However, by mid-2025, Volocopter was acquired by the Wanfeng Group in China to keep the company on track.

In the United States, Bristow Group, a global provider of vertical flight products such as helicopter transportation to energy customers and search and rescue aircraft, and Volocopter signed an agreement to develop passenger and cargo services for CAVs in the U.S. and U.K. Bristow also placed an order for two VoloCity EAVs with an option to purchase 78 more.

In late 2023, Volocopter conducted public demonstration flights over New York City, Las Vegas, Nevada, and Tampa, Florida, showing the public what an air taxi service would look like. Volocppter is but one example of the global nature of flying taxis with major cities around the world getting involved.

Volocopter is the standard rather than the exception in partnership activities by flying vehicle companies. For example, Overair in Santa Ana, California, partnered with the City of Arlington, Texas, for eVTOL operations starting at Arlington Municipal Airport and extending throughout the

DFW region. The EAV service would offer a direct connection to the city's Entertainment District, home to Six Flags Over Texas, Hurricane Harbor, four professional sports franchises, and numerous restaurants, shopping centers, museums, and live entertainment venues.

Overair and Dallas-Fort Worth International Airport also plan to bring eVTOL operations to the DFW Metroplex. Overair developed the Butterfly EAV, which can carry a pilot, and four passengers.

Uncrewed Flying

The ultimate end goal of numerous flying vehicle companies is to have their vehicles fly passengers without a pilot on board, commonly referred to as being uncrewed, By early 2024, only one company anywhere in the world was officially sanctioned to carry passengers, and that company now operates its flying vehicles pilot free.

EHang Holdings in China's Guangdong Province received the type certificate (TC) officially issued by the Civil Aviation Administration of China (CAAC), and in late 2023 received a standard airworthiness certificate (AC) from the CACC, making it the world's first unmanned eVTOL two-passenger vehicle cleared for commercial use.

EHang's passenger-carrying electric unmanned aerial vehicles (UAV) adhere to three main concepts: full backup safety design, unmanned operation, and centralized management enabled by the intelligent command- and-control center. The $410.000 EAV, with its intelligent autopilot, is designed to provide short-to-medium distance low-altitude trips. The idea is that unmanned flight technology can avoid failures caused by human errors, helping passengers enjoy a journey without worrying about control issues. If an emergency occurs during a trip, the command-and-control system can automatically select another route so the EAV can continue to the destination. EHang's UAVs utilize 4G/5G high-speed wireless network transmission links, enabling communication between the EAV and the command-and-control center, allowing remote control and real-time transmission of flight data.

The initial use of EHang's EH216-S vehicles is for aerial sightseeing at local scenic spots. Based on the strategic partnership agreement with EHang, the Hefei Municipal Government intends to facilitate the purchase of at least 100 units of EH216 series pilotless aerial vehicles from EHang, and/or provide financing support, amounting up to a total of $100 million. The Hefei government also plans to collaborate with EHang to expand aerial sightseeing flight routes and services to more scenic spots in Hefei, such as Swan Lake.

Other companies also plan to use EAVs for short trips. For example, Eve Air Mobility, which is backed by Embraer, and UrbanX Air in Miami are planning EAV flights by 2026. GlobalX agreed to purchase 200 EAVs from Eve Air and established UrbanX as a GlobalX subsidiary. Like other flying vehicle companies, Eve Air also has global ambitions.

The company and Flynas, a low-cost Saudi Arabia airline, teamed to launch eVTOL services in Riyadh and Jeddah. Another low-cost airline, Korean airline Jeju Air, and Eve Air are planning for EAV vehicles on Jeju Island. Eve Air also partnered with Hunch Mobility to launch a service in Bangalore, India.

Beyond Flying Taxis

Some companies that make flying taxis also are involved in the necessary infrastructure needed to service such vehicles. For example, air taxi developer Beta Technologies in Burlington, Vermont, received a $169 million loan to finance a net-zero final assembly production facility. The 188,500-square-foot facility is located at the Patrick Leahy Burlington International Airport, where Beta is building its eVTOL and CTOL (conventional takeoff and landing) fleet. Beta's Alia vertical and conventional takeoff vehicles are targeted for military and commercial use, including for medical, defense and passenger industries. The production facility, on a 40-acre site at the airport, could ultimately produce 300 EAVs. Beta Technologies also makes electric battery chargers that can charge its flying vehicle in less than one hour. Beta has installed charging systems at numerous airports, including at Eglin Air Force Base in Florida, where one of Beta's flying vehicles is being tested by the U.S. Air Force.

Beta, one of the strongest and broadest flying vehicle makers, also has a strategic partnership with GE, which made an equity investment of $300 million in Beta.

Flying taxi vehicles also are making their way into aerial tourism, such as with the EHang vehicles just discussed. Japanese EAV maker SkyDrive is selling eVTOLs to an MASC General Incorporated Association, a group that promotes aerial tourism in Setouchi Islands in western Japan. The SkyDrive vehicles are being manufactured at a Suzuki company plant, which could produce 100 flying vehicles a year.

Flying vehicle companies approach the market in numerous ways and with varying vehicles. For example, Lilium developed a vertical takeoff and landing, six-passenger jet, Crisalon created a patented stability system that controls the aircraft's movements in all directions, and Sirius Aviation aims to develop a hydrogen-powered EAV jet by 2028. However, like other startups mentioned earlier, Lilum also ran out of cash after spending $1.5 billion and ultimately went out of business, selling its patent portfolio to Archer Aviation $21 million.

8 PILOTLESS TRAVEL

The Uncrewed Future

The planned future of flying vehicles is to transport passengers between destinations without a pilot on board. This is commonly referred to as uncrewed urban air mobility (UAM), meaning vehicles flying without an onboard crew with UAM referring to lower-altitude operations within urban environments. There are numerous drivers of this coming approach, ranging from economic to technological. With thousands of low-altitude vehicles in the air, a challenge would be to find enough licensed pilots to operate them, along with all the associated costs. From a technological standpoint, many of the flying vehicles will be traveling on recurring routes, so mapping of air corridors is rather straightforward, Additionally, the entire air taxi business in general is aiming to keep consumer travel costs down, with some targeting air travel charges comparable or even less than car travel.

However, the move to uncrewed flights on a mass scale globally is down the road, with most certified flying vehicle being operated by onboard pilots. Meanwhile, others already are planning for uncrewed operations. For example, flying vehicles from EHang Holdings in China, the only certified passenger carrying flying vehicles, are flying pilotless, as discussed earlier in the book and as I witnessed at DriftX in Abu Dhabi. And advanced air mobility company Wisk, a subsidiary of Boeing, has taken an exclusively fully self-flying vehicle approach to the market. The San Francisco startup has conducted more than 1,700 test flights and its autonomous vehicles have flown in Long Beach, California. Wisk has been working on its fixed wing, self-flying vehicles for more than 13 years. For autonomous travel sometime in the future, Archer agreed to make Wisk its exclusive provider of autonomy technology for future variants of Archer's EAV, as mentioned earlier in the book.

Future Blueprint

Boeing and its subsidiaries Wisk and Aurora Flight Sciences laid out a detailed blueprint for pilotless passenger-carrying vehicles by the end of the decade. The Concept of Operations (ConOps) for Uncrewed Urban Air Mobility provides a roadmap to pilotless travel. "UAM operations will be conducted day and night in visual and instrument meteorological conditions," the document states. "The aircraft will utilize electric propulsion and employ a combination of onboard and ground-based systems to carry out functions currently executed by onboard pilots. Electric propulsion will eliminate direct carbon emissions, reduce noise, and reduce energy consumption. Uncrewed operations monitored and managed by multi-vehicle supervisors, in collaboration with automated systems, will reduce the cost of flight, thereby increasing access and affordability. The concept of operations for urban air mobility (UAM) provides a potential blueprint for global UAM operations. The ConOps targets safe initiation of uncrewed UAM passenger operations in the NAS (national air space) by the end of this decade, while providing a steppingstone to the goal of transitioning to high-throughput operations in the years that follow.

The context of the ConOps is the United States national airspace system (NAS) with the intention of conveying a blueprint for operations on a global scale. The key principles and assumptions in the ConOps are based on providing safe, affordable, everyday flight for everyone. The pilotless EAVs and complementary ground facilities would be tightly integrated into vehicle operations. The ground facilities would provide the usual airline fleet operations management and flight dispatch functions, and other essential functions currently performed by onboard pilots. Uncrewed EAVs would be managed by supervisors located in fleet operations enters, with supervisors responsible for monitoring assigned aircraft while not acting as a remote pilot.

Working in the System

One key of the ConOps as well as other views of future air travel is that the current aviation infrastructure would be leveraged rather than be replaced. The idea is for flying vehicles to be integrated with other commercial airline traffic along with coordination of takeoff and landing activities. The ConOps approach suggests pilots would remain on flying vehicles through 2028 with pilotless travel on many flying vehicle by 2032. The planning document accounts for EAVs that can carry four to six passengers over distances up to 69 miles at

1,200 to 4,000 feet above ground. EVAs would be equipped with an onboard flight operations management system with primary responsibility for safe execution of flight. The system would comprise a mission and vehicle management systems and be responsible for flight plan management and execution, air traffic control communications, vehicle flight, recording, and maintenance data capture. Travel routes would be planned by the fleet operation centers and filed with air traffic control for flow management and applicable separation. To deal with potential cyber threats, the uncrewed EAVs would require robust and traceable compliance with airworthiness security regulations.

In emergency situations, fleet operation centers can take remote control of the EAV, much like remotely flying a drone. For takeoffs and landings, the ConOps suggests vertiports could be located on top of skyscrapers and parking garages. "Vertihubs will host vertiport managers, ground management personnel, and base maintenance staff capable of performing heavy maintenance, repair, and overhaul on UAM aircraft," the ConOps states. "Because of the large space requirements and urban real estate constraints, vertihubs will typically be located in industrial or suburban areas. Vertistops, vertiports, and vertihubs may be co-located within airports (similar to general aviation terminals) or evolve from existing helistops or helipads (like those in Los Angeles and New York). Co-locating early vertiports will be advantageous due to the existing infrastructure, like terminals, with associated services for passengers and supporting CNS (communications, navigation, and surveillance) infrastructure."

The ConOps proposal identifies key principles for the integration of EAVs into the current airspace and transportation system:

- Existing infrastructure: Maximizing the use of the infrastructure, procedures,

and services provided by the existing air transportation system.

- Enabling regulation: New and revised regulations and operational policies will enable new operations and their integration into the NAS.
- High-fidelity flight planning: UAM operations will entail high levels of efficiency and predictability that will be attained by conformance to detailed flight plans with both spatial and time elements.
- Minimal conflicting traffic: Collaborative flight planning will enforce limits on traffic flow demand and will avoid intersecting flight paths. Flight plans will be end-to-end, from takeoff to landing, and will be approved and routinely executed in their entirety.

Booking flying vehicles such as air taxis will be an online activity with ground transportation such as ridesharing services connected as part of the process. Expect participating airlines to integrate flying vehicle bookings within their current reservation systems. There is much to establish before operations of fully autonomous EAVs can take place. Some of the requirements stated in the ConOps document are real-time airspace awareness, vertiport status, and wind maps, high-integrity digital communications, ATC voice communications via VHF radio relay, ATC data communications, precision navigation including in GPS-denied conditions, vertical autolanding, vertical auto-takeoff, and air to ground-based detect and avoidance systems. The rules and regulations for the automatic operations of urban air mobility will involve the entire aviation industry since all traditional airplane flying is affected. We discuss rules, regulations, and challenges in the next chapter.

9 RULES, REGUATIONS, ROADBLOCKS

Getting Flying Vehicles Off the Ground

While the development and testing of flying vehicles are well underway globally, there is a relatively lengthy and complex path ahead before the widescale operation of passenger-carrying flying vehicles can occur. The primary consideration is safety, with the protection of those in the vehicles and those on the ground paramount. There also are local considerations, such as where to locate takeoff and landing facilities, proximity to buildings such as schools, and ground traffic to and from the facility. Many of those factors will be taken up by local communities with local regulations. Coordination will be required for air traffic, manufacturers and operators, airport operators, port authorities, emergency management services, and federal, tribal, state, and local law enforcement organizations. Weather also could be an issue, with EAVs flying at low altitudes. "It's a challenge of people in the industry seeing the significance of weather," Don Berchoff, CEO of TruWeather Solutions, told me. As part of the low-altitude economy, TruWeather created a micro-weather detection system that could be used by EAV operators to determine whether a vehicle should fly. Besides weather factors and local ground considerations, federal regulators will be responsible for certifying that the vehicle meets certain standards and are totally safe to fly and carry passengers.

The good news is that the Federal Aviation Administration saw this coming and in mid-2023 issued the Advanced Air Mobility Implementation Plan, which the FAA defined as "an emerging aviation ecosystem that leverages new aircraft and an array of innovative technologies to provide the opportunity for more efficient, more sustainable, and more equitable options for transportation. AAM is a transportation system that moves people and property by air between two points in the United States using aircraft with advanced technologies, including electric aircraft, or electric vertical takeoff and landing (eVTOL) aircraft, in both controlled and uncontrolled airspace."

The FAA is responsible for bringing new technologies safely into aviation and its current challenge is to safely integrate AAM into the National Airspace System (NAS). The FAA is responsible for the operating rules, aircraft certification, and pilot certification. Numerous companies we interviewed while researching this book said they were working and coordinating their efforts with the FAA. At the federal level, the FAA is in charge of approving every necessary aspect to support AAM flights including the aircraft itself, the framework for operations, access to the airspace, operator training, infrastructure development, environmental impacts, and community engagement. For the near term ecosystem development, the FAA created a detailed approach called

Innovate28 (I28), a segment in the Advanced Air Mobility Implementation Plan. The FAA has teams that focus on AAM integration, including certification, airspace and air traffic management, infrastructure, environment, hazardous materials, safety, and community engagement, to ensure a coordinated approach.

Innovate for 2028

The Innovate28 (I28) initiative is designed to culminate with integrated AAM operations with manufacturers and/or operators flying between multiple origins and destinations at one or more locations in the United States by 2028. States the FAA: "For I28, AAM aircraft will be authorized for piloted operations and will transport passengers and/or cargo within the limits of the aircraft and certification regulations. The aircraft are expected to range in size from single passenger to larger occupancy shuttles, and employ new means of propulsion (e.g., electric motors, hydrogen fuel, hybrid designs). Many are capable of vertical takeoff and landing (VTOL) or short takeoff and landing (STOL) operations and quickly transition to fixed-wing operation after takeoff. It is assumed wake characteristics will be known, including the impacts of (air) wake from other aircraft on AAM aircraft and an AAM aircraft's own wake generation."

The FAA roadmap provides a scenario for flying vehicle operations:

1. The pilot follows established procedures for checking weather and NOTAMs (notice to air missions) for departure, en route, and destination, and files a flight plan if required. While passengers prepare to board the AAM aircraft, the pilot conducts aircraft walkarounds,

preparations, safety protocols, and departure checklists.

2. Departing in uncontrolled airspace, the pilot is responsible for adhering to appropriate rules governing flight in uncontrolled airspace. The pilot announces their departure intentions over a common radio frequency and maneuvers the aircraft to the takeoff location. After visually ensuring their departure area and path is clear, the pilot departs.
3. The pilot is aware of the requirements for flight in controlled airspace. The aircraft enters controlled airspace by means of two-way radio communication and the appropriate clearance from ATC. Published procedures or agreements (national, local, or signatory) reduce the need for ATC communications.
4. ATC issues instructions or clearances to provide separation and/or sequence the aircraft with other traffic. The pilot on board complies with instructions given by ATC or follows previously coordinated and approved instructions from the approving ATC facility (via published procedures or agreements).
5. ATC transfers control and communication from controller to controller as the aircraft transits different ATC sectors. After the pilot obtains information on destination runway(s) in use, weather, and other pertinent airport information, they start their approach to the landing site or follow a previously approved approach path. The ATC tower issues a clearance to land. The pilot may ask and be permitted to land at airport areas other than runways and taxiways.
6. The pilot on board completes landing checklists, their approach, and safe landing at a new, existing, or predetermined approved landing site. The aircraft is

maneuvered to the approved parking area for deplaning.

7. Passengers and crew follow established procedures for deplaning the aircraft. Following prescribed security procedures, the passengers exit the area or are directed to any further security screening required to enter the secure terminal area for connecting flights.

Before even getting to any final operational stage, the FAA must approve the actual flying vehicle. That certification process involves stages, including design approval, production approval, and airworthiness approval. Design approval involves reviewing and approving an aircraft's proposed design, including its systems, structures, and performance capabilities. Production approval ensures that an aircraft is built according to the design and meets the FAA's quality standards and airworthiness certification ensures that the aircraft is in a condition for safe operation and conforms to its approved design. Initial EAV operations are expected to start in 2025 with larger scale operations running by 2028. In late 2023, the FAA and the U.S. Air Force agreed to work tougher to integrate eVTOL vehicles into the National Air Space, so new air traffic control procedures are being studied as an additional area. In addition to requiring regulatory procedures, public acceptance and adoption will also be on the table.

The Naysayers

In every major technologically driven upheaval, there are many naysayers, a normal expectation when something comes along requiring a dramatic shift or change in behavior. When the world wide web, essentially the commercialization of the internet, came along, businesses around the world were somewhat blindsided. Many companies were slow to adapt as I found traveling the world doing presentation after presentation explaining why this was such a transformative big deal. I would tell them the net is the future and the transformation was not about just creating a website, which is what many only did at first. I would advise major businesses that the transformation coming was about totally transforming how their company would do business, that employees, customers, and businesses would all be networked together. While obvious now, business leaders and consumers at the time didn't grasp the significance of all the changes they and their businesses would be facing.

Another example is YouTube, which was created to take advantage of the new commercial web. Before the web, the idea of individuals creating videos that could be instantly shared with millions of people around the world with the touch of a button would not have been fathomed. And when the web and online sales started, nothing sold online was taxed. During that time, I appeared before the National Governors Association and suggested that they not start taxing online sales since the commercial web needed time to grow and for consumers and businesses to get used to using it. In the early 2000s, online sales took off, most notably since everything was tax free. That all changed by around 2021, when governments were allowed to tax online sales, which is now a common and accepted practice.

Expect the electric aerial vehicle revolution to follow a similar path. The marketplace will launch, go through some bumps and grinds, morph over time, and eventually move into the accepted-practice arena. It will seem foreign to many at first. When I speak with people outside the flying vehicle industry, they often do not believe flying vehicles are real or coming, do not think they would ever work within the current airspace, say they would never fly in one, or all of the above.

It's only natural to be skeptical of a major technology driven innovation at first, since the larger picture is not yet fully developed, understood, or seen. For example, at the beginning of the mobile revolution, started in earnest when Apple's iPhone hit the market in 2007, all the major airlines eventually created a mobile app. These initial apps contained only information about flight schedules for people to read on their mobile phones. If they wanted to book a flight, they had to call an airline or go online from a PC. However, the flight information listings were only the first step taken into the newly discovered mobile environment. It wasn't until United Airlines added flight booking capabilities to their app that the airline industry ultimately saw how they could leverage the new mobile technology. Fast forward to years later, and mobile devices are essential travel companions from flight booking and checking in to real-time flight tracking. There have been many naysayers with their noteworthy comments documented throughout history.

What They Predicted

- **1876**: "This 'telephone' has too many shortcomings to be seriously considered as a means of communication. The device is inherently of no value to us." — William Orton, President of Western Union, internal memo.
- **1889:** "Fooling around with alternating current (AC) is just a waste of time. Nobody will use it, ever." — Thomas Edison.
- **1903**: "The horse is here to stay, but the automobile is only a novelty, a fad." – Advice from the president of the

Michigan Savings Bank to Henry Ford's lawyer Horace Rackham not to invest in the Ford Motor Company.

- **1921**: "The wireless music box has no imaginable commercial value. Who would pay for a message sent to no one in particular?" — Associates of David Sarnoff responding to a call for investment in the radio.
- **1926**: "While theoretically and technically television may be feasible, commercially and financially it is an impossibility." — Lee DeForest, "Father of Radio" and a pioneer in the development of sound-on-film recording used for motion pictures.
- **1932:** "There is not the slightest indication that nuclear energy will ever be obtainable. It would mean that the atom would have to be shattered at will." — Albert Einstein.
- **1943**: "I think there is a world market for maybe five computers. – IBM Chairman Thomas Watson
- **1946:** "Television won't be able to hold on to any market it captures after the first six months. People will soon get tired of staring at a plywood box every night." — Darryl Zanuck, film producer, co-founder of 20th Century Fox.
- **1949:** "Where a calculator on the ENIAC is equipped with 18,000 vacuum tubes and weighs 30 tons, computers of the future may have only 1,000 vacuum tubes and perhaps weigh one and a half tons." — Popular Mechanics.
- **1957:** "I have traveled the length and breadth of this country and talked with the best people, and I can assure you that data processing is a fad that won't last out the year." — Editor of Prentice Hall business books.

- **1959:** "The world potential market for copying machines is 5,000 at most." IBM to the eventual founders of Xerox.
- **1961:** "There is practically no chance communications space satellites will be used to provide better telephone, telegraph, television, or radio service inside the United States" — T.A.M. Craven, Federal Communications Commission (FCC) commissioner.
- **1966:** "Remote shopping, while entirely feasible, will certainly flop. It has no chance of success." – Time Magazine
- **1977:** "There is no reason for any individual to have a computer in their home." — Ken Olsen, president of Digital Equipment Corp.
- **1981:** "Cellular phones will absolutely not replace local wire systems." — Marty Cooper, inventor.
- **1989:** "We will never make a 32-bit operating system." — Bill Gates, co-founder and chairman of Microsoft.
- **1992:** "The idea of a personal communicator in every pocket is a "pipe dream driven by greed." — Andy Grove, then CEO of Intel.
- **1995:** "I predict the Internet will soon go spectacularly supernova and in 1996 catastrophically collapse." — Robert Metcalfe, founder of 3Com, inventor of Ethernet.
- **1997:** "Apple is already dead." -- Nathan Myhrvold, former Microsoft CTO.
- **1999:** "The era of the PC is over"-- Louis Gerstner, Chairman and CEO of IBM
- **2003:** "The subscription model of buying music is bankrupt. I think you could make available the Second Coming in a subscription model, and it might not be successful." — Steve Jobs in Rolling Stone

- **2004:** "Two years from now, spam will be solved." -- Bill Gates
- **2005:** "There's just not that many videos I want to watch." Steve Checn, CTO and co-founder of YouTube.
- **2006:** "Everyone's always asking me when Apple will come out with a cell phone. My answer is, 'probably never.'" David Pogue, The New York Times.

2007: "There's no chance that the iPhone is going to get any significant market share." — Steve Ballmer, Microsoft CEO.

We expect there will be a quote relating to flying vehicles that can be added to the famous quotes of the past one day. Perhaps one of the most fitting comments about the creation of flying vehicles can be the one from the late Apple founder Steve Jobs: "It's really hard to design products by focus groups. A lot of times, people don't know what they want until you show it to them."

The world is about to see flying vehicles.

Appendix

APPENDIX A

Glossary -- The Language of Flying Vehicles

AAM – Advanced Air Mobility
AAV – Autonomous Aerial Vehicle
ADS-B -- Automatic Dependent Surveillance
ATC – Air traffic control
BEV – Battery electric vehicles
C2 – Commands and control
CNS – Communication, navigation and surveillance
CTOL – Conventional takeoff and landing
DAA –Detect and avoid
EASA – European Union Aviation Safety Agency
EAV – Electric Aerial Vehicle
ELZ – Emergency landing zone
EVTOL – Electric Vertical Takeoff and Landing
FAA – Federal Aviation Administration

FBO – Fixed base operator
FOC – Fleet operation center
GCAA – General Civil Aviation Authority of UAE
ICAO – International Civil Aviation Organization
IFR – Instrument flight rules
NAS -- National Air Space
PAV – Personal Air Vehicle
PAVC—Personal air vehicle certified
UAM – Urban Air Mobility
UAMV – Urban Air Mobility Vehicle
UAS – Unmanned aircraft systems
UAV – Unmanned aerial vehicle
STOL—Short takeoff and landing vehicle
SVO -- Simplified vehicle operator
TC – Type Certification
VAS – Vertiport automation system
VCA – VTOL-Capable Aircraft
Vertiport – Takeoff and landing facility for eVTOLs
VFR – Visual flight rules
VTOL – Vertical takeoff and lading

APPENDIX B

Federal Aviation Administration (FAA) PART 103

Ultralight VEHICLES

Authority:

Subpart A—General

103.1 Applicability.

This part prescribes rules governing the operation of ultralight vehicles in the United States. For the purposes of this part, an ultralight vehicle is a vehicle that:

(a) Is used or intended to be used for manned operation in the air by a single occupant;

(b) Is used or intended to be used for recreation or sport purposes only;

(c) Does not have any U.S. or foreign airworthiness certificate; and

(d) If unpowered, weighs less than 155 pounds; or

(e) If powered:

(1) Weighs less than 254 pounds empty weight, excluding floats and safety devices which are intended for deployment in a potentially catastrophic situation;

(2) Has a fuel capacity not exceeding 5 U.S. gallons;

(3) Is not capable of more than 55 knots calibrated airspeed at full power in level flight; and

(4) Has a power-off stall speed which does not exceed 24 knots calibrated airspeed.

103.3 Inspection requirements.

(a) Any person operating an ultralight vehicle under this part shall, upon request, allow the Administrator, or his designee, to inspect the vehicle to determine the applicability of this part.

(b) The pilot or operator of an ultralight vehicle must, upon request of the Administrator, furnish satisfactory evidence that the vehicle is subject only to the provisions of this part.

103.5 Waivers.

No person may conduct operations that require a deviation from this part except under a written waiver issued by the Administrator.

103.7 Certification and registration.

(a) Notwithstanding any other section pertaining to certification of aircraft or their parts or equipment, ultralight vehicles and their component parts and equipment are not required to meet the airworthiness certification standards specified for aircraft or to have certificates of airworthiness.

(b) Notwithstanding any other section pertaining to airman certification, operators of ultralight vehicles are not required to meet any aeronautical knowledge, age, or experience requirements to operate those vehicles or to have airman or medical certificates.

(c) Notwithstanding any other section pertaining to registration and marking of aircraft, ultralight vehicles are not required to be registered or to bear markings of any type.

Subpart B—Operating Rules

103.9 Hazardous operations.

(a) No person may operate any ultralight vehicle in a manner that creates a hazard to other persons or property.

(b) No person may allow an object to be dropped from an ultralight vehicle if such action creates a hazard to other persons or property.

103.11 Daylight operations.

(a) No person may operate an ultralight vehicle except between the hours of sunrise and sunset.

(b) Notwithstanding paragraph a of this section, ultralight vehicles may be operated during the twilight periods 30 minutes before official sunrise and 30 minutes after official sunset or, in Alaska, during the period of civil twilight as defined in the Air Almanac, if:

(1) The vehicle is equipped with an operating anticollision light visible for at least 3 statute miles; and

(2) All operations are conducted in uncontrolled airspace.

103.13 Operation near aircraft; right-of-way rules.

(a) Each person operating an ultralight vehicle shall maintain vigilance so as to see and avoid aircraft and shall yield the right-of-way to all aircraft.

(b) No person may operate an ultralight vehicle in a manner that creates a collision hazard with respect to any aircraft.

(c) Powered ultralights shall yield the right-of-way to unpowered ultralights.

103.15 Operations over congested areas.

No person may operate an ultralight vehicle over any congested area of a city, town, or settlement, or over any open air assembly of persons.

103.17 Operations in certain airspace.

No person may operate an ultralight vehicle within Class A, Class B, Class C, or Class D airspace or within the lateral boundaries of the surface area of Class E airspace designated for an airport unless that person has prior authorization from the ATC facility having jurisdiction over that airspace.

103.19 Operations in prohibited or restricted areas.

No person may operate an ultralight vehicle in prohibited or restricted areas unless that person has permission from the using or controlling agency, as appropriate.

103.20 Flight restrictions in the proximity of certain areas designated by notice to airmen.

No person may operate an ultralight vehicle in areas designated in a Notice to Airmen under, unless authorized by:

(a) Air Traffic Control (ATC); or

(b) A Flight Standards Certificate of Waiver or Authorization issued for the demonstration or event.

[Doc. No. FAA–2000–8274, Sept. 11, 2001]

103.21 Visual reference with the surface.

No person may operate an ultralight vehicle except by visual reference with the surface.

103.23 Flight visibility and cloud clearance requirements.

No person may operate an ultralight vehicle when the flight visibility or distance from clouds is less than that in the table found below. All operations in Class A, Class B, Class C, and Class D airspace or Class E airspace designated for an airport must receive prior ATC authorization as required.

APPENDIX C

Following are the key portions of the FAA Part 103 Advisory providing details and guidance for compliance with FAA Part 103, the regulations for personal aerial vehicles.

This advisory circular provides guidance to the operators of ultralights in the United States. It discusses the elements which make up the definition of ultralight vehicles for the purposes of operating under Federal Aviation Regulation (FAR) Part 103. It also discusses when an ultralight must be operated as an aircraft under the regulations applicable to certificated aircraft.

Background

The number of ultralight vehicles and participants in the various aspects of this sport has increased dramatically in recent years. All indications are that this growth will continue. The presence of these vehicles in the national airspace has become a factor to be considered in assuring the safety of all users of the airspace.

On October 4, 1982, a new regulation (Part 103) applicable to the operation of ultralight vehicles became effective. This regulation defines those vehicles which may be operated as "ultralight vehicles" and provides operating rules which parallel those applicable to certificated aircraft. The Federal Aviation Regulations regarding aircraft certification pilot certification, and aircraft registration are not applicable to ultralight vehicles or their operators.

Ultralight vehicle operations may only be conducted as sport or recreational activity. The operators of these vehicles are responsible for assessing the risks involved and assuring their own personal safety. The rules in Part 103 are intended to assure the safety of those not involved in the sport, including persons and property on the surface and other users of the airspace. The ultralight community is encouraged to adopt good operating practices and programs in order to avoid more extensive regulation by the Federal Aviation Administration (FAA).

Definitions. For the purpose of this advisory circular, the following definitions apply:

Ultralight Vehicle. This term refers to ultralights meeting the applicability for operations under Part 103.

Recognized Technical Standards Committee. This term refers to a group of at least three persons technically qualified to determine whether a given ultralight meets the requirements for operations under Part 103, as follows:

It is recognized by a national pilot representative organization.

It is comprised of persons not directly associated with the manufacture and/or sale of the make of ultralight king inspected, and

It conducts its review and documents the findings in accordance with the guidance provided in this circular.

What does this mean for the person who wants to fly under Part 103?

You are responsible for your personal safety. Certificated aircraft are designed, flight tested, manufactured, maintained, and operated under federal regulations intended to provide an aircraft of consistent performance, controllability, structural integrity, and maintenance. An ultralight vehicle is not subject to federal aircraft certification and maintenance standards. This means that the costs of purchasing and maintaining an ultralight vehicle may be considerably less than the purchase of a certificated aircraft. There is no assurance that a particular ultralight vehicle will have consistent performance, controllability, structural integrity, or maintenance. Your safety, and potentially that of others, depends on your adherence to good operation and maintenance practices. This includes proper preflight techniques, operation of the vehicle within the manufacturer% recommended flight envelope, operation only in safe weather conditions, and providing safety devices in anticipation of emergencies. Part 103 is based on the assumption that any individual who elects to fly an ultralight vehicle has assessed the dangers involved and assumes personal responsibility for his/her safety.

You are Limited to single-occupant operations.

Part 103 is based on the single-occupant concept; operation by an individual who has assumed all responsibility for his/her personal safety. Pilots of ultralight vehicles subject to Part 103 are not required to have training or previous experience prior to the operation of these vehicles. You should consider receiving adequate training prior to participation.

You are limited to recreation and sport purposes. Operations for any other purpose are not authorized under the applicability of Part 103.

You are limited as necessary for the safety of other persons and property.

Part 103 consists of operating rules which were determined necessary for the-safety of other users of the airspace and persons on the surface. These rules were developed in consideration of the capabilities of the vehicles and their pilots. It is your responsibility to know, understand and comply with these rules. Ignorance of the regulations pertaining to the activities you pursue is not an acceptable excuse for violating those regulations.

You are Responsible for the Future Direction the Federal Government Takes With Respect to Ultralight Vehicles. The actions of the ultralight community will affect the direction Government takes in future regulations. The safety record of ultralight vehicles will be the foremost factor in determining the need for further regulations.

FAA contact points

The FAA will provide clarification of particular subject areas, information, and assistance pertaining to the operations of ultralight vehicles through the following contacts:

Flight Standards Field Offices.

Flight Standards District Offices (FSDOs), General Aviation District Offices, and Manufacturing and Inspection

District Offices are the FAA field offices where information and assistance are available regarding the operation of ultralight vehicles, acceptable methods of complying with Part 103 requirements, and compliance with other regulations should it become necessary to operate an ultralight as a. certificated aircraft.

Air Traffic Control Facilities.

FAA Air Traffic Control facilities are located throughout the United States and maintain jurisdiction over the use of the controlled airspace in their particular area. To obtain authorization to operate from or into the airspace designated in 103., contact must be made with the controlling facilities.

Flight Service Stations. These facilities provide operational information to pilots, such as weather briefings, advisory information regarding the status of facilities, etc., and are the most accessible of the FAA points of contact, They can provide additional information regarding how to reach the other points of contacts mentioned here.

Airports District Offices.

These offices inspect airports certificated under Part 139 of the FARs to determine whether an airport is safe for public use. Persons wanting to establish new airports or flight parks, or operate ultralight vehicles from Federally-funded airports, may contact these offices for assistance,

Wha is an ultralight vehicle?

SCOPE AND CONTENTS. This section discusses the elements contained in 103 which make up the definition of an 'ultralight vehicle" and the proper way to assure that Part 103 applies.

APPLICABILITY OF PART 103.

Probably the single most critical determination which must be made is whether or not your vehicle and the operations you have planned are permitted under Part 103. The fact that you are operating a vehicle which is called or advertised as a "powered ultralight," 'hang glider,'' 'or "hang balloon" is not an assurance that it can be operated as an ultralight vehicle under Part 103. There are a number of elements contained in 103 which make up the definition of the "ultralight vehicle." If you fail to meet any one of the elements, you may not operate under Part 103. Any operations conducted without meeting all of the elements are subject to all aircraft certification, pilot certification, equipment requirements, and aircraft operating rules applicable to the particular operation.

The FAA realizes that it is possible to design an ultralight which, on paper, meets the requirements of 103, but in reality, does not. However, the designers, manufacturers of the kits, and builders are not responsible to the FAA for meeting those requirements. Operators of ultralights should bear in mind that they are responsible for meeting 103 during each flight. The FAA will hold the operator of a given flight responsible if it is later determined that the ultralight did not meet the applicability for operations under Part 103.Be wary of any designs which are advertised as meeting the requirements for use as an ultralight chicle, yet provide for performance or other design innovations which are not in concert with any element of 103. The FAA may inspect any ultralight which appears, by design or performance, to not comply with 103.

If the FM Determines Your ultralight was not eligible for Operation as an Ultralight Vehicle.

If your ultralight does not meet 103, it must be operated in accordance with applicable aircraft regulations. You will be subject to enforcement action ($1000 civil penalty for each violation) for each operation of that aircraft.

Elements making up the definition of an ultralight vehicle

Single Occupancy. An ultralight cannot be operated under Part 103 if there is more than one occupant or if it has provisions for more than one occupant.

Sport or Recreational Purposes Only. An ultralight cannot be operated under Part 103 if it is operated for purposes other than sport or recreation or if it is equipped for other uses.

No Airworthiness Certificate. An ultralight cannot be operated under Part 103 if it has been issued a current U.S. or foreign airworthiness certificate.

Unpowered Vehicles. An unpowered ultralight cannot be operated under Part 103 if it weighs 155 pounds or more. Balloons and gliders are unpowered vehicles.

Powered Vehicles. A powered ultralight cannot be operated under Part 103 when it has an empty weight of 254 pounds or more; has a fuel capacity exceeding 5 U.S. gallons; is capable of more than 55 knots (63 miles per hour) airspeed at full power in level flight; and has a power-off stall speed which exceeds 24 knots (28 miles per hour).

13 l SINGLE OCCUPANT.

The Rationale for Allowing Single-Occupant Operations Only.

One aspect of the rationale for allowing ultralight vehicles to operate under special rules which do not require pilot and aircraft certification is the single-occupant limitation. The assumption is made that a person who elects to operate an uncertificated vehicle alone is aware of the risks involved. This assumption does not necessarily hold true for a passenger. Because the pilot qualifications for ultralight vehicle operations are not Federally controlled or monitored, the single-occupant requirement is a necessary component oi the continuation of the policies and regulations which allow the operation of ultralight vehicles free from many of the restrictions imposed on the operation of certificated aircraft.

Guidelines Regarding Seating Arrangements Which Should be Considered when Purchasing or Operating an Ultralight vehicle.

Any provisions for more than one occupant automatically disqualify an ultralight for operations under Part 103.

Some powered ultralights were originally manufactured with bench or love seats with only one seatbelt but have been advertised as two-place in the ultralight periodicals, they are not eligible for operations under Part 103. While no maximum width standards for the size of a single seat have been established at this time, most manufacturers are providing seats which have a width of 18 to 22 inches. Any seat notably wider than 22 inches raises a question as to whether the ultralight is intended for single occupancy.

An ultralight with provisions for more than one occupant can only be operated as a certificated aircraft, even when occupied by only one person. In addition to the previously stated aircraft certification and registration requirements the pilot must hold a medical certificate and at least a student pilot certificate with the proper endorsements for solo operations. At least one occupant during two-occupant operations must hold at least a private pilot certificate.

Two-place Ultralight Operations under Part 103.

The AOPA Air Safety Foundation, Experimental Aircraft Association, and the United States Hang Gliding Association have been granted exemptions from the appliable aircraft regulations to authorize use of two-place ultralights under Part 103 for limited training purposes and for certain hang glider operations, except as authorized by exemption, no person may operate an ultralight under Part 103 with m-an one occupant.

Recreation and sport purposes only

The Rationale for Allowing Recreation and Sport Operations Under Part 103.

In combination with the single-occupant requirement, the limitation to recreation to sport operations only is the basis for allowing ultralight vehicle operations under minimum regulations. The reason for allowing the operation of these vehicles without requiring aircraft and pilot certification is that this activity is a sport* generally conducted away from concentrations of population and aircraft operations.

Determining whether a Particular Operation is for Recreation and Sport purposes.

There are several considerations that are necessary in determining whether a given operation is conducted for recreation or sport purposes:

Is the flight undertaken to accomplish some task, such as patrolling a fence line or advertising a product? If so, Part 103 is not applicable.

Is the ultralight equipped with attachments or edifications for the attachments of modifications, such as banner towing or agricultural spraying? If so, Part 103 does not apply.

Is the pilot advertising his/her services to perform any task using an ultralight? If so, Part 103 does not apply.

Is the pilot receiving any form of compensation for the performance of a task using an ultralight vehicle? If so, Part 103 does not apply.

Examples of Operations Which are Clearly Not for Sport or Recreational Purposes.,

Aerial Advertising. Part 103 does not apply to operations that include the towing of banners and the use of loudspeakers, programmed light chains, smoke writing, dropping leaflets, and advertising on wings; nor does it apply to the use of interchangeable parts with different business advertisements or flying specific patterns to achieve maximum public visibility.

Aerial Application. Part 103 does not apply to operations that include using an ultralight to perform aerial application of any substance intended for plant nourishment, soil treatment, propagation of plant life or pest control. An ultralight with an experimental certificate as an amateur aircraft could be used to perform this function under specific, limited circumstances. Paragraph 35b provides more detail on this subject.

Aerial Surveying and Patrolling, Patrolling powerlines, waterways, highways, suburbs, etc., does not come under Part 103. The conduct of these activities in an ultralight must be in compliance with applicable aircraft regulations as outlined in paragraph 34. Local, state, or Federal government entities may operate an ultralight as a public aircraft.

Carrying parcels for hire.

Examples of Situations Involving Money or Some Other Form of compensation allowable under the recreation and sport limitation.

Rental of Ultralight Vehicles. Renting an ultralight vehicle to another person is permissible.

Receiving a Purse or Prize. Persons participating in sport or competitive events involving the use of ultralights are not prohibited from receiving money or some other form of compensation in recognition of their performance,

Authoring Books About Ultralights.

Persons are not prohibited from flying ultralights and then authoring books about their experiences, for which they ultimately receive compensation.

Receiving Discount on Purchase of an Ultralight.

There is no prohibition which would prevent you from taking advantage of any discount on the price of an ultralight a company might offer where its logo or name appears on a portion of the vehicle. You cannot, however, enter into any agreement which might specify the location, number, or pattern of flights contingent on the receipt of that discount. Any operation under such an agreement could not be conducted under Part 103.

Participation in Airshows and Events. You may participate in airshows and other special events where persons are charged for viewing those events, so long as you receive no compensation for your participation. This does not hold true where you stand to benefit directly from the proceeds as the organizer producer of the event.

Appendix

Index

Chuck Martin has been a leading digital pioneer for more than two decades. He is the founder of Net Future Institute, which focuses on disruptive business strategies and tactics for the hyper-connected world of people, places, and things. He is the host of the worldwide Podcast "The Voices of the Internet of Things with Chuck Martin" and former Editor of the AI & IoT Daily at MediaPost.

Chuck was early and accurate in predicting the web revolution, on target in predicting the size and scope of the mobile revolution, and is now forecasting a major transformation in electric, aerial mobility. He is the author of numerous business books, including *New York Times* Business Bestseller *The Digital Estate, Net Future* and *The Third Screen.* Chuck is an international speaker, has been named #1 in Internet of Things Top 10 Influencers, and IoT Thought Leaders to Watch. He is on the forefront of research exploring the roles of flying vehicles, which he has been tracking and writing about for years.

www.ingramcontent.com/pod-product-compliance
Lightning Source LLC
Chambersburg PA
CBHW031934190326
41519CB00007B/521
9780976327349